Hardware to Code

How Software Is Transforming
the Automotive Industry

Mukesh Sharma

To my family, whose unwavering love and support have been the foundation of my journey. You have been my constant source of strength and inspiration, and I am forever grateful for the sacrifices you've made.

To my friends, who have shared in my triumphs and challenges over the years. Your encouragement and camaraderie have been invaluable, and I cherish the memories we've created together.

To my leaders, past and present, who have challenged me, guided me, and provided opportunities for growth. Your leadership and mentorship have shaped me into the professional I am today.

And to my colleagues, with whom I've had the privilege of collaborating. Your insights, dedication, and teamwork have enriched my journey and contributed immensely to my learning.

This book is a reflection of all that I've learned over the past 26 years, and I owe so much of it to each of you.

CONTENTS

Part 1

How Software is Transforming the Automotive Industry

Part 2
The Business of Software Defined Vehicles

FOREWORD

A Collective Voice to Set the Stage

Throughout my journey in the automotive industry, I've had the privilege of working closely with several inspiring leaders—visionaries who have shaped not only the trajectory of the industry but also my own thinking. Their influence has been instrumental in the way I view the transformation from hardware-driven engineering to software-defined mobility.

That's why, instead of including a single voice, I chose to gather perspectives from multiple leaders whose work I deeply admire. Their voices reflect the diversity of thought, experience, and conviction that is shaping the future of the automotive world.

There are many more voices I hope to include in future editions, as this transformation is far from over. But I am truly grateful to those who took the time to share their views in this edition.

What better way to set the tone than to open the book with insights from those who are actively defining the next chapter of mobility?

My heartfelt thanks to Jack Cheng, I.V. Rao, C.V. Raman, Ming Jing, and Karthikeyan Natarajan for contributing their perspectives to Hardware to Code. Their reflections offer readers a rich, grounded entry into the chapters that follow—and I hope their words will spark the same curiosity and conviction that inspired this book.

Foreword & Insights by Jack Cheng

As the automotive industry undergoes one of the most profound transformations in history, the rise of Software-Defined Vehicles (SDVs) stands at the very heart of this revolution. The convergence of artificial intelligence, data, and electrification is not only reshaping how vehicles are designed, built, and experienced, but it's also redefining the future of mobility itself.

I have had the privilege of being part of this journey from the early days at Nio, where we focused on creating a smarter, more sustainable electric vehicle ecosystem, to founding M Mobility and helping pioneer an open-source, collaborative approach for the future of the EV industry. My time at XPT, Ford & Fiat Chrysler further reinforced my belief in the power of technology to drive innovation across the automotive landscape. Through these experiences, I've seen first-hand how software is not just an add-on to vehicles but the very foundation upon which the future of mobility is built.

This book captures the pivotal moment in automotive history—where SDVs are no longer a distant vision but a reality shaping the way we drive, live, and interact with the world. It is a powerful resource that demystifies the complexities of SDVs and provides invaluable insights into their potential, challenges, and the opportunities they present across industries. The in-depth exploration of this emerging technology is a crucial guide for both industry professionals and enthusiasts who are keen to understand the intricacies of SDVs and their role in transforming the automotive ecosystem.

As someone who has long believed in the power of collaboration and open-source innovation, I particularly appreciate how your book explores not only the technology but also the ecosystem that supports it. The work you've done in highlighting the role of key players, such as OEMs and startups, is invaluable. It's essential that we build a cohesive, interconnected network to drive SDV development forward, and your insights into this will play a key role in shaping the future of mobility.

Looking ahead, we are only at the beginning of what promises to be a transformative journey. The next few years will be critical in determining how SDVs will evolve, the infrastructure that will support them, and the regulatory frameworks that will ensure their safe and responsible integration into society. Your book serves as both a roadmap and a call to action for those of us who are working tirelessly to shape this future.

I wholeheartedly believe that the insights within these pages will inspire the next generation of innovators, engineers, and leaders in the SDV space. The journey is just beginning, and I am excited to see how your work will continue to contribute to this groundbreaking shift in the world of transportation.

Best regards,
Jack Cheng
Co-Founder of Nio,Former CEO of XPT,
Ex CEO of MIH, Founder of M Mobility

Foreword & Insights by I.V. Rao

Global auto industry is witnessing transformative technological changes to address GHG emissions and make vehicles more interactive for users as an extension of office environment, and Indian auto sector is no exception.

The Indian government has also initiated very aggressive measures for energy security and improves Air quality and meets Net Zero economy goals

As a result the global auto industry is at a defining crossroads. Software-Defined Vehicles (SDVs) take center stage—particularly in advanced markets like the United States and China. There are lots of changes in how vehicles are designed, developed, and experienced. China has made remarkable progress in adopting SDVs, driven by its bold investments, digital ecosystem, and policy alignment.

While software is undoubtedly becoming the soul of modern vehicles, it is important not to overlook the body—the hardware. The role of traditional Tier-1 suppliers will certainly evolve, and their relationship with OEMs is already being redefined. But this isn't the end of hardware innovation—it's realignment. Smart hardware will coexist with intelligent software to deliver next-generation mobility.

In the Indian context, software will open new doors for enhancing customer experience, safety, and connectivity.

However, the India is one of few growing markets with very low penetration of vehicles in all segments. With India's economic growth ambitions, auto industry has high growth potential. Also Indian customers are very price sensitive and majority are entry level customers. Hence buyer's priorities still revolve around fundamentals—affordability, fuel efficiency, and driving comfort. And the younger generation values connectivity etc for continuity in work on move. SDVs in India will be adopted for high end segments and may slowly percolate down.

What I find truly compelling about this book is that it brings a global lens, especially from China, where the author, Mukesh, has spent years immersed in the fast-moving, highly innovative auto landscape. His deep experience is evident across these pages. Hardware to Code is not just a technical exploration—it's a strategic mirror for industry leaders, startups, and policymakers, especially in India, to reflect upon where we are and where we must go. I believe this book will serve as an eye-opener and an invaluable guide for all those shaping the future of mobility."

Best regards,

IV Rao

Former Executive Director – Engineering & R&D,

Maruti Suzuki India Ltd.

FOREWORD & INSIGHTS BY KARTHIKEYAN NATARAJAN

The automotive industry is undergoing one of the most profound shifts in its 100-year history. At the heart of this transformation lies a singular truth: code now defines the car, the next century horizon

For decades, automotive excellence was measured in horsepower, torque, fuel efficiency, and ride comfort. But today, those metrics are being replaced—or at least rebalanced—by lines of software, intelligent architecture, and cloud connectivity. We are witnessing the rise of the Software-Defined Vehicle: a platform that is less a mechanical product and more a rolling supercomputer on wheels

This isn't about just adding electronics or ADAS safety features. This is about rethinking the very DNA of automotive development led by software architecture From zonal architectures to centralized compute platforms, from real-time operating systems to AI inference engines—every layer of the vehicle is being re-coded for intelligence, safety, adaptability, and over-the-air evolution.

The real challenge—and opportunity—is orchestration. Building a car is no longer an exercise in mechanical, electronics and software integration, but in software synchronization & integration . You need automotive-grade platforms that can operate with real-time determinism and safety, but also integrate cloud services, edge AI, and digital twins. You need teams that understand embedded

C and deep learning, systems engineering and DevOps pipelines, functional safety and user experience—all in the same conversation.

This is not an easy pivot for the traditional auto industry as the technology innovation 'S curves' go through changes, the new winners emerge. It requires rewiring mindsets, supply chains, and business models. It demands that companies think like software enterprises, not just hardware OEMs. Those who fail to adapt will be left behind—not because of cost, but because of irrelevance. I've seen this evolution up close. For over a decade, I've worked at the intersection of embedded systems, AI, IoT, and next-gen mobility.

This is why Hardware to Code is timely and important. Mukesh Sharma brings a practitioner's lens to the most disruptive transition in mobility's history. His insights are grounded in real work, real markets, and real experience. For anyone building, investing in, or simply trying to understand the future of vehicles—this book offers a powerful guide.

Let this book be your ignition point.
Let software be your engine.
Let vision be your GPS.

Best regards,
Karthikeyan Natarajan
Former CEO & Executive Director, Cyient

FOREWORD & INSIGHTS BY CV RAMAN

I have been associated with the Auto Industry for more than 4 decades. The industry has over the years gone through transformation in a steady manner which has kept up with the customer needs & regulatory framework. The controls have moved from mechanical to electro-mechanical to fully electronic. The machines and devices have shrunk and have been able to deliver with more precision, efficiency and speed.

However we have seen an unprecedented disruption in the last 7-10 years. This got kick started with the growth of the connectivity & the rapid adoption of internet based services. The Environment & Safety regulation coupled with the rising aspirations of the customers for connected, safe & convenience features in their vehicles has pushed the Engineers & Product Development engineers to innovate. Mobile connectivity & IOT based solutions have been increasingly adopted and the customer desires the vehicles as an extension of their homes. Connected, Autonomous, Shared and Electric technologies are being propagated which has grown the hardware and software & compute power demands.

This complexity of hardware and software and the move to unlock value added services is making Industry to move to Software Defined Vehicles. This gives flexibility for customization, software updates, enhance functionalities & also rectify potential issues that may come up. With AI & generative AI, the software complexity may get simplified. The challenge is need for enhanced cyber security, increase in reliability & validation, increased redundancy and high development & maintenance costs. The technology cost absorption may be possible in the higher segments.

Making lighter and scalable platforms would be the next challenge or opportunity for the Industry which can be applied across various segments. I am sure Mukesh's book would unravel some of the challenges & opportunities in this realm. Wishing him & the book all the success which will surely will become a must read for current & future auto professionals!

Best regards,
CV Raman
Executive Director Maruti Suzuki India Ltd

Foreword & Insights by Ming Jing

The automotive industry is undergoing a shift, and at the heart of this transformation is the emergence of Software-Defined Vehicles (SDVs). As someone who has spent over two decades immersed in automotive technology, and the last 13 years closely connected with the author of this book, I can confidently say that we are witnessing a redefinition of roles, responsibilities, and relationships across the entire automotive value chain.

Traditionally, Tier-1 suppliers have played a crucial role in shaping the driving experience—delivering high-quality hardware and tightly integrated software components. However, as OEMs increasingly backward-integrate and bring core software development in-house, the balance of power is shifting. Software is no longer a black box delivered by suppliers; it is becoming a central pillar of OEM strategy, product differentiation, and customer engagement.

This change is not without its challenges. Tier-1s must now rethink their positioning—no longer just as component providers, but as co-creators and ecosystem orchestrators. Strategic collaboration is becoming essential, not only with OEMs but also with semiconductor giants and cloud hyperscalers. Chipmakers are moving up the value chain, offering more integrated solutions, including software stacks and domain controllers. Meanwhile, cloud providers are powering the digital backbones for over-the-air updates, data lakes, AI training, and DevOps pipelines.

That's why this book is so timely and important. It provides a clear, structured lens through which to view the SDV transformation—not just from a technical standpoint, but also from an ecosystem and business model perspective. The author skillfully bridges the gaps between software architecture, platform economics, and the shifting alliances among OEMs, Tier-1s, chipmakers, and cloud providers.

As we look to the next era of mobility—one that is autonomous, connected, electrified, and software-defined—collaboration will be key. No single player can win this race alone. I believe this book will serve as a guidepost for industry professionals, helping them navigate this disruptive journey with clarity and purpose.

I'm honored to contribute to this work and to support its vision for a more open, intelligent, and collaborative automotive future.

Warm regards,
Ming Jing
VP and Head of Segment Asia, Architecture
& Networking Solutions Business Area
Continental Automotive

PREFACE

When I first shared my idea in November 2023 about writing a book on automotive engineering and business, my wife Nidhi responded with a candid question: "So many books are published every day—what difference will yours make?" That moment sparked something in me. It challenged me not just to write a book, but to write one that would truly stand out, one that would add value, especially for those navigating the rapidly evolving world of mobility.

Initially, I envisioned a purely technical deep dive into Software-Defined Vehicles (SDVs). But as the manuscript evolved, so did my perspective. I realized that over the last decade, my journey had expanded far beyond engineering—I had also lived through boardrooms, strategic alliances, customer pain points, and market disruptions. I had transformed from a hands-on technologist to a business leader immersed in shaping the future of the industry.

That realization led me to structure the book into two interconnected parts:

Part 1 explores how software is fundamentally transforming the engineering and architecture of vehicles.

Part 2 dives into how this software revolution is reshaping the business models, supply chains, and ecosystems that define the automotive world.

When I began my career as a vehicle testing engineer in 1998, the industry was defined by mechanical excellence. By the time I shifted into a business leadership role in 2010, software was already creeping into the driver's seat. But it was around 2013 that I saw the tide turn—code began to shape cars more than hardware ever did, and the transformation hasn't slowed since.

The title Hardware to Code captures that very shift. This book is a blend of curated insights, countless hours of reading, first-hand industry experience, and AI-assisted research. It is written to inform, provoke thought, and engage not just engineers, but decision-makers, strategists, and enthusiasts who care about where mobility is headed.

I see this book as just the beginning. There is far more to write—entire threads I've intentionally left untouched for future exploration. But I believe this book will spark dialogue, challenge conventions, and offer a holistic lens on the technological and business revolutions underway.

And perhaps a decade from now, I'll be able to turn to Nidhi and say, "You were right to question me—but I believe this did make a meaningful contribution."

To spark your curiosity and help you navigate this fast-evolving world of Software-Defined Vehicles (SDVs), I've included a set of thought-provoking questions at the beginning of each chapter summary. These questions are not just rhetorical—they are the real-world challenges and strategic dilemmas that the automotive industry is grappling with today. By presenting each chapter through a lens of inquiry, my intention is to guide you into the core ideas, help you frame your understanding, and make your reading journey more engaging and purposeful. Consider these questions as your travel guide through the complex landscape where code meets metal, and where innovation is shifting gears faster than ever before.

Part 1: How Software Is Transforming the Automotive Industry

1. The Evolution of the Global Automotive Industry
   How did the automotive industry evolve from a mechanical marvel to a digital ecosystem? What were the key milestones that shaped this journey?

2. The Tipping Point – From Hardware to Code Why is the automotive industry shifting from a hardware-driven approach to a software-centric model?
   What opportunities and challenges arise from this transformation?

3. Understanding Software-Defined Vehicles What defines a Software-Defined Vehicle (SDV)?
   How do SDVs revolutionize the way cars are designed, manufactured, and experienced?

4. Architecture & Operating Systems
   How are new vehicle architectures and operating systems reshaping the automotive ecosystem? Why is centralized computing essential for SDVs?

5. Open Source Development: The Blockbuster for Software-Defined Vehicles?
   Can open-source software accelerate innovation in the automotive industry? What role it will play in the rise of SDVs?

6. Regulations & Cyber Security Challenges
   What are the key regulatory and cybersecurity challenges that automakers face in the age of SDVs? How can these hurdles be overcome?

7. Impact of Generative AI

   How is Generative AI transforming the automotive industry? Can AI revolutionize vehicle design, prototyping, and even autonomous driving systems?

8. Chip-to-Cloud Strategies

   What is the significance of chip-to-cloud connectivity for SDVs? How do these strategies enhance vehicle performance, security, and user experience?

9. Role of Cloud Providers

   Why are cloud providers becoming indispensable to the automotive industry? How do partnerships with hyperscalers unlock the potential of SDVs?

10. How Electromobility Is Transforming by Software-Defined Vehicles

    How is software accelerating the adoption and efficiency of electric vehicles? What role do SDVs play in redefining electromobility?

11. How Software-Defined Vehicles Drive Cost Optimization for Carmakers

    Can SDVs help automakers reduce costs across production, operations, and maintenance? How do over-the-air updates drive efficiency?

Part 2: The Business of Software-Defined Vehicles

12. China's Automotive Journey

    How has China emerged as a global leader in the SDV and EV revolution? What lessons can the world learn from China's aggressive innovation strategies?

19. The Road Ahead: A Snapshot in Time
    What lies ahead for the automotive industry as it navigates the intersection of hardware, software, and mobility trends? Can SDVs drive the industry toward sustainable, intelligent mobility solutions?

To conclude the book, Chapters 20 and 21 take a practical turn—offering case studies and deep insights from key stakeholders within the Software-Defined Vehicle (SDV) ecosystem. These chapters are designed to bridge theory with practice. Chapter 20 presents real-life case studies that explore how different players—OEMs, Tier 1 suppliers, startups, and technology providers—are navigating the SDV transformation. Chapter 21 features an in-depth interview with Dr. Edward Tse, Founder and CEO of Gao Feng Advisory, whose visionary insights shed light on China's unique role in the global mobility shift and the strategic imperatives facing the industry. Together, these final chapters ground the broader themes of the book in real-world experience and foresight, leaving you with a comprehensive and nuanced view of the road ahead.

ACKNOWLEDGEMENTS

As I reflect on my 26-year journey across the automotive and technology landscape, I am filled with gratitude for the remarkable leaders who shaped my thinking, mentored my growth, and inspired the perspectives shared in this book.

I would like to sincerely thank Gopalkrishnan and Jayant Chakraborty at Tata Motors for laying the foundation of my professional journey and nurturing my early passion for automotive innovation.

At Maruti Suzuki, I had the privilege of learning under the guidance of visionaries like IV Rao, CV Raman, P. Panda, Deepak Sawkar, Mizuno-san, and Shigeki Suzuki-san. Their blend of technical excellence and leadership profoundly influenced my understanding of product engineering and the evolution of vehicles.

My journey in Tech Mahindra was enriched by the leadership and support of CP Gurnani, Rohit Gandhi, Raghavendra Tripathi, Amitava Ghosh, Manoj Chugh, Sujit Baksi, Anuj Bhalla, Harshvendra Soin and Mohit Joshi. Each of them contributed to shaping my global outlook, strategic thinking, and leadership journey, especially as I navigated the dynamic Greater China region.

To all these leaders—whether I worked directly or indirectly — thank you for your inspiration, trust, and the opportunities you gave me to grow, challenge myself, and contribute.

This book is as much a reflection of your influence as it is of my experiences.

The list is so long that probably I cannot write names of all in this version of the book, however I can say to all, whoever reads it or whoever comes to know about it in my known circle that everyone has postively motivated me to pen my thoughts.

Thank you to all my well wishers

PART 1

How Software is Transforming the Automotive Industry

ONE

The Evolution of the Global Automotive Industry

The global automotive industry has witnessed more transformation in the past two decades than in the entire century before it. From the combustion-powered carriages of the early 1900s to today's AI-powered, software-defined vehicles, the industry has not only evolved—it has been redefined. What began as a mechanical engineering marvel has now turned into a digital platform economy on wheels. This chapter in brief traces that fascinating journey—through the mechanical age, mass production era, globalization wave, electrification push, and now, the age of intelligence and code.

1. The Mechanical Age (1885–1920s): Birth of the Automobile

The roots of the automotive industry lie in the late 19th century, with Karl Benz's Motorwagen (1885) often credited as the first true automobile. In this era, automobiles were luxury products—handcrafted, expensive, and unreliable. The innovation lay primarily in mechanical engineering: internal combustion engines, transmissions, and suspensions. These machines were complex, and every unit was almost bespoke.

2. The Assembly Line Revolution (1920s–1950s): Scale and Standardization

The real inflection point came when Henry Ford introduced the moving assembly line in 1913. This single change democratized the automobile. Cars like the Model T were now affordable, and production scaled exponentially. This era saw rapid standardization, growth of automotive supply chains, and the birth of iconic brands—General Motors, Chrysler, Volkswagen, Toyota. Design, engineering, and manufacturing became the core of competitive advantage.

3. The Globalization Wave (1960s–1990s): Competition and Quality

By the 1960s, Japanese automakers like Toyota, Honda, and Nissan introduced lean manufacturing and Kaizen (continuous improvement) philosophies, which transformed quality and efficiency benchmarks. Western OEMs began facing intense competition. The global automotive map was redrawn—Germany, Japan, the U.S., and South Korea emerged as major players. This was also the era of global supply chains, platform-sharing across geographies, and vehicle segmentation (SUVs, hatchbacks, sedans) based on consumer demographics.

4. The Digital Decade (2000s–2010s): Electronics and Embedded Systems

The early 2000s ushered in a quiet revolution: electronics and software began creeping into vehicles. Anti-lock braking systems (ABS), electronic stability programs (ESP), infotainment systems, GPS navigation—all powered by embedded software and ECUs (Electronic Control Units). A modern car began to house 80–100 ECUs, each controlling a specific function.

However, the software was still tightly coupled to hardware, often developed in silos, and updates required dealership visits. The car was becoming a computer—but not yet an intelligent one.

5. The Electric and Mobility Shift (2010s–2020s): Powertrain Disruption

Two parallel trends hit the automotive world in the 2010s:

- Electrification: Spearheaded by Tesla, EVs transitioned from quirky experiments to viable products. Battery technology improved, costs dropped, and environmental pressures mounted.
- Mobility-as-a-Service (MaaS): Ride-sharing apps (Uber, Didi, Ola) changed car ownership dynamics in urban areas.

OEMs were forced to rethink the vehicle powertrain, customer interface, and business model. Governments joined in, with aggressive EV mandates and urban clean-air policies.

6. The Software-Defined Vehicle Era (2020s–Present): From Hardware to Code

We now enter perhaps the most disruptive phase in automotive history. Cars are being redefined not by engines or gearboxes, but by software stacks, data architectures, and connectivity.

Key transformations:

- Centralized computing architectures: Replacing dozens of ECUs with powerful domain and zonal controllers.
- Software-Defined Vehicles (SDVs): Where features, performance, and even monetization are controlled via over-the-air (OTA) updates.
- Autonomous driving and ADAS: Pushing vehicles toward self-awareness and decision-making.
- Vehicle-as-a-Platform (VaaP): With in-vehicle apps, voice assistants, and subscriptions.

- Digital twins & AI-driven development: Accelerating prototyping, testing, and customization.

This shift is not just technical—it's economic and strategic. Tesla, with fewer models but deep software integration, has a valuation higher than the next 5 OEMs combined. Players like BYD, Xiaomi, Rivian, and Lucid are entering with tech-first strategies. Traditional OEMs are rushing to form software divisions, invest in cloud-native development, and hire AI engineers. Valuation now hinges not on how many cars you build, but on how smart, adaptable, and upgradeable those cars are.

7. Future Outlook: Intelligence, Integration, and Ecosystems

The future automotive company will be as much a software company as it is a manufacturer. Key trends shaping the next decade include:

- Generative AI: Designing, testing, and even co-piloting cars.
- Smart infrastructure: V2X (Vehicle-to-Everything) systems for urban mobility.
- Digital services: Cars generating more revenue post-sale than at point-of-sale.
- Open ecosystems: COVESA, Eclipse SDV, and more enabling collaborative innovation.

8. China: The New Epicenter of Automotive Innovation

In the last decade, China has not only caught up but is now setting the pace for global automotive innovation. It is no longer just the world's largest auto market—it is becoming the industry's innovation lab. Nowhere is this more evident than in the electric and software-defined vehicle space.

EV Leadership: Chinese brands like BYD, NIO, XPeng, and Li Auto are now global forces. BYD has surpassed Tesla in EV volumes, and Chinese battery giants like CATL dominate global supply chains.

SDV Acceleration: China's ecosystem—rich in software talent, fast-moving startups, and AI innovation—is pushing SDV development at unmatched speeds.

Reverse Technology Transfer: For the first time in automotive history, the flow of innovation is reversing. Western OEMs are importing EV and SDV platforms and components from Chinese suppliers. Partnerships and joint ventures are being formed to learn from China's pace and scalability.

Smart Infrastructure & V2X: Chinese cities are rapidly piloting smart roads, 5G-based vehicle-to-infrastructure (V2I) systems, and autonomous shuttle networks.

This aggressive paradigm shift led by China is reshaping the traditional East-to-West flow of technology. China is no longer a factory—it is now the ideation and engineering hub for the mobility future.

Conclusion: Reinventing the Wheel

The story of the automotive industry is one of relentless reinvention. What began as a mechanical machine to transport humans is now evolving into a connected, intelligent, and adaptive platform. From steel frames to silicon chips, from horsepower to processing power, from oil to data—the automotive world has turned its engine inside out.And at the heart of this shift is a single, powerful idea: the future of mobility lies not just in wheels—but in code.

TWO

THE TIPPING POINT – FROM HARDWARE TO CODE

The transformation of vehicles into intelligent, software-defined platforms is no longer a futuristic concept—it is an industrial reality. Over the past decade, the automotive industry has undergone a fundamental shift, moving from a hardware-centric development model to one that prioritizes software as the new engine of innovation. This shift is not merely technical—it represents a deep cultural and operational evolution across OEMs, suppliers, and developers.

At the heart of this evolution is the concept of the Software-Defined Vehicle (SDV). Unlike traditional vehicles, where functionality was predominantly tied to mechanical systems and embedded hardware, SDVs place software at the center of differentiation. The legacy model of vehicles containing hundreds of ECUs operating in silos is giving way to centralized computing architectures, domain controllers, and service-oriented platforms. These enable real-time feature updates, cross-domain coordination, and dynamic scalability—fundamentals required to manage the projected 300 million lines of code per vehicle by 2030.

This transformation is driven by necessity and possibility. Consumers now expect cars to behave like smartphones—offering continuous feature updates, personalized experiences, and seamless integration. For the industry, that means adopting new development models rooted in DevOps, containerization, and OTA (Over-the-

Air) services. Frameworks like AUTOSAR Adaptive, Android Automotive, and middleware-based architectures are emerging as critical enablers. The rise of open-source ecosystems like Eclipse SDV and alliances like COVESA illustrates a growing consensus: innovation can no longer happen in isolation.

Yet, SDVs bring unique complexities. Unlike phones, cars carry lives and responsibilities—demanding rigorous attention to safety, lifecycle durability, and real-time performance. This balance is pushing the industry to adopt virtualization layers, hardware abstraction, and hypervisors that decouple software from silicon while maintaining performance integrity. Tesla's model of pre-installed hardware activated through software updates sets a new precedent, blending product lifecycle with continuous innovation and monetization—via subscription-based features and performance unlocks.

The implications ripple across the entire ecosystem. OEMs must now rethink their organizational structures, transitioning from function-based teams to integrated software-product organizations. Tier-1 suppliers can no longer rely on black-box models—they must embrace transparency, co-innovation, and modularity. Even chipmakers are stepping up, moving from Tier-2 to Tier-0.5 status by offering full-stack, customizable solutions directly to OEMs.

This chapter marks a crucial inflection point in automotive history. As mechanical constraints give way to digital possibilities, the core question shifts from "What can the vehicle do?" to "What can the code enable?" In this new world, hardware becomes the foundation—but software becomes the soul. This is the evolution—from hardware to code.

THREE

Understanding Software Defined Vehicles

In the dynamic realm of the automotive industry, the advent of software-defined vehicles (SDVs) marks a pivotal juncture. These vehicles, characterized by their dependence on advanced software systems to govern various functions, signify not only a technological advancement but also a strategic transition that requires meticulous planning from industry stakeholders.

Software is revolutionizing car capabilities while also posing development challenges for automotive players. Adopting an end-to-end approach that integrates individual software components into a comprehensive platform can enhance functionality and reduce complexity.

Today, generative AI can greatly enhance SDV development and operation by automatically generating artifacts such as test cases, architecture models and software source code. Equally important, SDVs require many specialized technologies from different providers which is why ecosystem collaboration plays an important part in the SDV architecture.Ultimately, every component in the architecture plays a well-defined role in ensuring the best possible experience for vehicle drivers and passengers, solidifying the SDV as the next evolution of the automotive industry.

Today's vehicles, traditionally hardware-centric, are swiftly evolving into software-defined transportation platforms. Recent automotive innovations, such as intuitive infotainment, autonomous driving capabilities, and electrification, rely less on mechanical prowess and more on software sophistication. This rapid transformation has left automotive original equipment manufacturers (OEMs) and other industry players grappling to keep pace. Moreover, the substantial cost associated with integrating and upgrading features that consumers increasingly demand, such as high-end onboard assistants and advanced driver-assistance systems (ADAS), presents a daunting challenge.

Understanding Software Defined Vehicles

At its essence, a software-defined vehicle transcends mere transportation; it embodies a sophisticated computing platform on wheels. Leveraging a blend of sensors, processors, connectivity, and intelligent algorithms, SDVs possess the capability to process vast amounts of data in real-time, adapt to evolving conditions, and deliver a seamless, personalized user experience.

Despite the evident significance of software in vehicle performance, the development of automotive software modules often occurs independently. OEMs may develop some modules in-house, while others are procured from suppliers or emerge from strategic alliances or joint ventures. Subsequently, OEMs or their tier-one suppliers endeavor to integrate these modules into a proprietary platform.

A typical modern vehicle likely comprises a software architecture consisting of numerous domains, encompassing hundreds of functional components within the car and across the cloud. These domains encompass various facets, including infotainment, ADAS, mapping, telematics, and third-party applications. In constructing this architecture, OEMs interact with a myriad of software providers to develop diverse capabilities, resulting in a multitude of development languages, operating systems, and software architectures within vehicles. This fragmented approach is prevalent among industry leaders, as no single software platform can adequately address all cross-system requirements.

Many automotive companies leverage foundational code for their software stacks, including operating systems and critical middleware, from other industries. By doing so, they significantly shorten development timelines and reduce costs, as creating original code poses significant challenges. For instance, infotainment systems for upcoming vehicles often derive from the smartphone industry, with automotive adaptations of mobile operating systems becoming increasingly common. Similarly, software for ADAS originally drew from aerospace applications and manufacturing automation, with real-time operating systems from semiconductor manufacturers and embedded software companies gaining traction.

While individual software modules have facilitated notable automotive advancements over the past decade, the industry is entering a new era where seamless integration among multiple vehicle subsystems is paramount. For example, the active suspension systems in several luxury vehicles necessitate real-time collaboration among ADAS cameras, powertrain sensors, and chassis actuators— three distinct domains with separate software architectures, operating systems, and middleware.

Despite the critical need for integration among vehicle systems to unlock novel use cases, companies lack an end-to-end platform to seamlessly connect these components. Consequently, OEMs and tier-one suppliers grapple with the formidable task of interface control and integration, leading to significant challenges in development, security, and performance.

The development challenge facing the automotive industry is becoming increasingly tangible. As the prevalence of software within vehicles continues to rise, so too does the complexity of integrating various systems into a cohesive whole. Automotive players are struggling to keep pace with these demands, exemplified by the significant time and effort required to develop modern infotainment systems. These systems now take upwards of three years to develop, with a multitude of software engineers contributing to each iteration. A substantial portion of this effort—typically ranging from 30 to 50 percent—is dedicated to integration, given the extensive network of suppliers involved in development. Furthermore, changes to one software module often necessitate extensive rework, as these systems are not always backward compatible and require substantial redevelopment every few years to remain current with new features and performance standards.

Adding to the complexity is the projection that software complexity will nearly triple over the next decade. This poses a significant challenge for OEMs and suppliers, as their productivity is not increasing at a rate sufficient to sustain innovation. With the gap between complexity and productivity widening, OEMs and tier-one suppliers are expected to face a severe talent shortage and a substantial increase in development costs. While OEMs are acknowledging the productivity problem, they recognize that there is no quick fix. With talent in short supply, simply allocating additional resources to the problem is not a viable solution. Instead, they must focus on streamlining software development by reducing complexity.

Moreover, the surge in complexity is giving rise to a myriad of new security risks for connected vehicles. Advances in in-vehicle networking have created links between previously independent electronic domains, such as infotainment, ADAS, and powertrain.

While these connections facilitate the introduction of new features, they also provide a pathway for attacks to spread throughout a vehicle. Software vulnerabilities in one system can be exploited to gain access to other systems, posing significant security concerns. Coordination among developers working on different software stacks is lacking, making it difficult to align updates and patches across modules. This challenge is exacerbated by the increasing number of potential "attack surfaces" as the prevalence of connected and autonomous driving systems rises.

Connectivity has emerged as the primary performance driver in today's cars, surpassing hardware in importance. Seamless connectivity across domains is essential for enabling new features and innovations. In internal-combustion engines, software has facilitated recent advancements such as rapid stop-start technology and variable valve timing to enhance efficiency. For electric vehicles (EVs), software plays an even more critical role, particularly in managing trade-offs between performance and range. For example, EVs must contend with issues such as reduced range in extreme temperatures due to the absence of excess heat and power generated by internal combustion engines. Efficient software systems are essential for coordinating across domains and managing these trade-offs effectively.

Internal connectivity within vehicles is equally crucial. While software has enabled advanced infotainment capabilities, new challenges have emerged, including clunky interfaces and disconnects between different components. Drivers are often confronted with lag-filled interfaces that are slow to respond to inputs, leading to frustration and dissatisfaction. These issues not only detract from the driving experience but also fall short of consumer expectations for streamlined, user-friendly interfaces. Addressing these challenges will be essential for automotive players to meet evolving consumer demands and deliver superior in-vehicle experiences.

A better path forward

Envision a world where software integration throughout a vehicle is seamless from end to end. In this scenario, a primary operating system, robust and adaptable enough to encompass major systems within the vehicle, along with software modules developed on a shared code base, would serve as the foundation for this integration. Such a construct could alleviate many of the challenges prevalent in today's fragmented ecosystem.

First and foremost, this integrated approach would enable automotive players to address performance issues stemming from disparate operating systems and disjointed sets of code. Cross-domain interfaces could become significantly simpler, facilitating direct communication between different systems without the need for translation that often compromises efficiency and introduces delays. Additionally, security measures could be greatly enhanced, with a unified solution capable of monitoring the entire code base, rather than just a handful of interfaces. Moreover, development productivity could see substantial improvement, as OEMs could introduce new cross-domain features over time without extensive software rewrites, leveraging a common code base across the vehicle.

End-to-end software solutions could also pave the way for a markedly improved human-machine interface. Foundational user-experience elements could be easily shared across vehicle domains, ensuring consistency in behavior, appearance, and feel across output methods such as center screens, instrument clusters, and heads-up displays. This standardized approach could enhance safety by enabling real-time notifications to trigger immediate visual, auditory, and tactile feedback throughout the cockpit, prompting drivers to take action. As interior design preferences evolve and user interfaces are adapted for different vehicle models and tiers, the sensory experience could be adjusted with minimal development effort.

Furthermore, end-to-end software platforms could facilitate dynamic resource sharing, leading to reduced overall hardware costs while enabling the addition of new capabilities over time. OEMs are moving towards an approach where in-vehicle communications align with an Ethernet standard, creating a real-time, high-bandwidth information highway that links major systems and allows for dynamic resource sharing across the vehicle. A unified operating system could allocate processing power as needed, dynamically enhancing performance for critical systems. For instance, during high-complexity "edge cases," more resources could be allocated to advanced driver-assistance systems (ADAS) to ensure swift responses. This optimization could lead to cost savings in hardware and enable the seamless integration of additional processing capacity to support future applications.

Initiating Change: Paths for Automotive Industry Stakeholders

The shift towards a common architecture capable of hosting code for each major system has already taken place in other industries. For example, drone developers have transitioned from individual operating systems for flight control, wireless communication, and cameras to a unified platform that integrates code across the device, simplifying development and enabling new cross-domain functionalities. Similarly, leading smartphone manufacturers have improved functionality by closely integrating code across hardware components and reusing a single software architecture across various devices, reducing the need for redesign.

Similar trends are emerging in the automotive industry, with select tech-focused players taking preliminary steps towards integrating software elements into combined platforms. For instance:

A prominent smartphone software provider, established in infotainment, is expanding into hypervisors to enable safety-critical driver communication required for ADAS.

A provider of real-time ADAS operating systems is diversifying into infotainment offerings to create a safety-critical solution for driver notifications.

A major tech company with a cloud-services platform is broadening its scope into applications, offering a tailored software suite for automobiles.

As these trends gain momentum, it's evident that a strategic shift towards end-to-end software integration holds the potential to revolutionize the automotive industry, ushering in a new era of innovation and efficiency.

The Strategic Imperative

For automotive companies, embracing software-defined vehicles represents more than just keeping up with technological trends—it's a strategic imperative for future success. Here's why:

1. Competitive Advantage: In a fiercely competitive market, differentiation is key. By investing in software-defined vehicles, automakers can differentiate their offerings with advanced features, enhanced connectivity, and superior user experiences that resonate with consumers.

2. Future-Proofing: The automotive industry is undergoing a profound transformation, with trends such as electrification, autonomous driving, and mobility-as-a-service reshaping the competitive landscape. Software-defined vehicles offer a flexible platform that can adapt to these changes, ensuring relevance and longevity in an uncertain future.

3. Monetization Opportunities: Software-defined vehicles open up new avenues for revenue generation beyond traditional vehicle sales. From subscription-based services and over-the-air updates to data monetization and ecosystem partnerships, SDVs enable innovative business models that can drive incremental revenue streams.

4. Customer Engagement: Today's consumers expect seamless connectivity and personalized experiences in every aspect of their lives, including their vehicles. By embracing software-defined vehicles, automakers can deepen their engagement with customers through features such as advanced infotainment systems, predictive maintenance, and smart mobility services.

Key Considerations for Success

While the benefits of software-defined vehicles are clear, realizing their full potential requires a strategic approach. Here are some key considerations for automakers embarking on the journey towards SDVs:

1. Investment in Talent: Building software-defined vehicles requires a diverse set of skills, including software engineering, data analytics, cybersecurity, and user experience design. Automakers must invest in acquiring and developing talent with expertise in these areas to drive innovation and execution.

2. Partnerships and Ecosystem Collaboration: No company can do it alone. Automakers must cultivate partnerships and collaborate with technology companies, software vendors, and other ecosystem players to leverage complementary strengths, accelerate innovation, and create value for customers.

3. Regulatory Compliance and Cybersecurity: As vehicles become increasingly connected and reliant on software, regulatory compliance and cybersecurity become paramount. Automakers must prioritize compliance with applicable regulations and implement robust cybersecurity measures to protect vehicles and customer data from cyber threats.

4. Customer-Centric Design: The success of software-defined vehicles hinges on their ability to deliver value to customers. Automakers must adopt a customer-centric approach to design, focusing on understanding customer needs, preferences, and pain points to create compelling and intuitive user experiences.

5. Agility and Adaptability: The automotive industry is evolving rapidly, with new technologies, business models, and consumer preferences emerging constantly. Automakers must be agile and adaptable, able to iterate quickly, experiment with new ideas, and pivot as needed to stay ahead of the curve.

Cutting Edge features of SDVs (till the time this book was under preparation, the world of SDV is changing so fast that many more features will evolve soon)

Software-driven vehicles offer a plethora of new features that were not available in traditional cars. Here are some distinct features from various automakers:

1. Toyota:

Predictive Maintenance: Software-driven vehicles can utilize data analytics to predict maintenance needs based on real-time vehicle diagnostics, helping to prevent breakdowns and optimize performance.

Advanced Driver Assistance Systems (ADAS): Toyota vehicles may offer features like lane-keeping assistance, adaptive cruise control, and automatic emergency braking to enhance safety and driving convenience.

2. Nio:

Battery Swapping Technology: Nio's software-driven vehicles may incorporate battery swapping capabilities, allowing drivers to quickly replace depleted batteries with fully charged ones at designated stations, thereby reducing charging times and increasing convenience.

In-Car AI Assistant: Nio vehicles might feature an intelligent AI assistant that can provide personalized recommendations, control various vehicle functions, and interact with passengers using natural language processing.

3. Xpeng:

Autonomous Parking: Xpeng's software-driven vehicles could offer autonomous parking capabilities, allowing drivers to remotely park their vehicles using a smartphone app or voice commands.

Over-the-Air (OTA) Updates: Xpeng vehicles may receive regular OTA updates to enhance performance, add new features, and improve cybersecurity without the need for physical service appointments.

4. Huawei:

Huawei HiCar Integration: Huawei's software-driven vehicles might seamlessly integrate with the Huawei HiCar ecosystem, allowing users to access smartphone apps, navigation, and other services directly from the vehicle's infotainment system.

5G Connectivity: Huawei vehicles could leverage 5G connectivity for faster internet speeds, improved navigation accuracy, and enhanced vehicle-to-everything (V2X) communication, enabling advanced safety and convenience features.

5. BMW:

Gesture Control: BMW's software-driven vehicles may offer gesture control technology, allowing drivers to interact with the infotainment system, adjust settings, and answer calls with simple hand gestures.

Personalized Driver Profiles: BMW vehicles might support personalized driver profiles that automatically adjust seat positions, climate settings, and entertainment preferences based on individual driver preferences.

6. General Motors:

Super Cruise: General Motors' software-driven vehicles could feature Super Cruise, an advanced driver assistance system that offers hands-free driving on compatible highways, allowing drivers to relax and enjoy the ride while the vehicle maintains lane position and speed.

Enhanced Vehicle Connectivity: General Motors vehicles may offer enhanced connectivity features, such as remote vehicle monitoring, diagnostics, and control via smartphone apps, enabling greater convenience and peace of mind for drivers.

7. BYD:

Vehicle-to-Grid (V2G) Integration: BYD's software-driven vehicles might integrate with V2G technology, allowing them to discharge stored energy back to the grid during peak demand periods, potentially reducing electricity costs for owners and supporting renewable energy integration.

Advanced Energy Management: BYD vehicles could feature advanced energy management systems that optimize battery usage, charging schedules, and energy regeneration to maximize range and efficiency.

8. Volkswagen (VW):

Augmented Reality Head-Up Display (AR HUD): VW's software-driven vehicles may offer an AR HUD that projects vital information, such as navigation instructions, speed, and collision warnings, onto the windshield, enhancing situational awareness without distracting drivers.

Intelligent Voice Control: VW vehicles might feature intelligent voice control systems that can understand natural language commands and interact with various vehicle functions, allowing drivers to stay focused on the road while accessing infotainment and navigation features.

9. Daimler:

Mercedes-Benz User Experience (MBUX): Daimler's software-driven vehicles could feature MBUX, an advanced infotainment system with AI-powered voice recognition, natural language understanding, and predictive suggestions for navigation, entertainment, and comfort settings.

Energizing Comfort Control: Daimler vehicles may offer Energizing Comfort Control, a feature that synchronizes climate control, ambient lighting, seat massage, and multimedia systems to create personalized wellness experiences tailored to driver and passenger preferences.

10. Geely Zeekr:

Advanced Connectivity: Geely Zeekr's software-driven vehicles may offer advanced connectivity features, including seamless integration with smart home devices, voice assistants, and cloud services, allowing users to control home appliances, access multimedia content, and manage schedules from the vehicle's infotainment system.

Geely Smart Cockpit: Geely Zeekr vehicles might feature the Geely Smart Cockpit, a state-of-the-art interface that combines digital displays, touchscreens, and gesture control to provide intuitive access to vehicle functions, navigation, and entertainment options.

11. Nissan:

ProPILOT Assist: Nissan's software-driven vehicles may feature ProPILOT Assist, an advanced driver assistance system that combines adaptive cruise control and lane-keeping assist to provide semi-autonomous driving capabilities on highways and in heavy traffic conditions.

NissanConnect: Nissan's proprietary infotainment system, NissanConnect, offers features such as smartphone integration, remote vehicle control, and emergency assistance, enhancing connectivity and convenience for drivers.

12. Honda:

Honda Sensing: Honda's software-driven vehicles may come equipped with Honda Sensing, a suite of safety and driver-assistive technologies, including collision mitigation braking, lane-keeping assist, and adaptive cruise control, to enhance driver safety and confidence on the road.

HondaLink: Honda's connected car platform, HondaLink, provides features such as remote start, vehicle diagnostics, and roadside assistance, allowing users to stay connected to their vehicles from their smartphones.

13. Stellantis:

Uconnect: Stellantis' Uconnect infotainment system offers advanced features such as navigation, entertainment, and smartphone integration, providing seamless connectivity and entertainment options for drivers and passengers.

Stellantis Connected Services: Stellantis offers connected services such as remote vehicle access, vehicle tracking, and emergency assistance, enhancing safety, security, and convenience for drivers of their software-driven vehicles.

14. Tesla:

Autopilot: Tesla's software-driven vehicles are equipped with Autopilot, an advanced driver assistance system that enables features such as adaptive cruise control, lane-keeping assist, and automatic lane changes on highways.

Full Self-Driving (FSD): Tesla offers Full Self-Driving capability as an optional upgrade, which includes features such as Navigate on Autopilot, Auto Lane Change, and Autopark, allowing for more autonomous driving capabilities in various road conditions.

Over-the-Air (OTA) Updates: Tesla vehicles receive regular OTA software updates that introduce new features, improvements, and bug fixes, ensuring that the vehicles remain up-to-date with the latest advancements in autonomous driving and infotainment technology.

15. Renault:

EASY CONNECT: Renault offers its EASY CONNECT services, providing advanced connectivity features such as remote vehicle monitoring, navigation with real-time traffic updates, and over-the-air software updates. These services enhance the overall user experience and keep Renault vehicles up-to-date with the latest features and improvements.

Renault EASY LINK: The Renault EASY LINK infotainment system provides seamless integration with smartphones, offering access to music streaming, navigation apps, and hands-free calling. The intuitive interface and voice recognition technology make it easy for drivers to stay connected while focusing on the road.

It is important for me to connect the dots of SDV to achieveing carbon neutrality. The road to carbon neutrality will be a significant step for SDV journey.Achieving carbon neutrality is a crucial goal in our fight against climate change, and software-defined vehicles offer promising solutions to help us get there. These innovative vehicles, powered by advanced software systems, can significantly reduce carbon emissions in several ways:

Efficient Power Management: Software-defined vehicles optimize how they use power, whether they're electric or hybrid. This means they can make better use of energy, reducing the amount of carbon emitted during travel.

Optimized Driving Patterns: Smart software algorithms analyze how we drive and adjust the vehicle's performance to be more fuel-efficient. This includes things like optimizing acceleration, deceleration, and finding the most efficient routes.

Regenerative Braking: These vehicles can capture energy during braking and reuse it, rather than wasting it as heat. This not only saves energy but also cuts down on emissions.

Vehicle-to-Grid Integration: Software-defined vehicles can connect to smart grids, allowing them to store and distribute energy. This helps balance energy supply and demand, reducing the need for carbon-intensive power generation.

Optimized Maintenance: Smart software can predict when parts might fail, helping us fix them before they cause emissions-heavy breakdowns. Plus, it keeps vehicles running smoothly, which means fewer emissions overall.

Eco-Driving Assistance: Some software-defined vehicles offer tips to drivers on how to drive more efficiently. This includes advice on things like accelerating gently and maintaining a steady speed, all of which can cut down on emissions.

Carbon Offset Tracking: These vehicles can keep track of how much carbon they emit, helping us understand anda offset our environmental impact.

Reduced Lifecycle Carbon Footprint: Smart software can also make manufacturing more efficient, using fewer resources and producing less waste. Plus, it helps us recycle vehicles more effectively, further reducing their environmental impact.

Fleet Optimization: For businesses with vehicle fleets, software-defined vehicles can help optimize routes and schedules, reducing unnecessary driving and emissions.

Integration with Sustainable Mobility: Software-defined vehicles can work with other sustainable transportation options, like ride-sharing and public transit, helping reduce the number of cars on the road and emissions overall.

In conclusion, software-defined vehicles have the potential to make a big difference in our efforts to achieve carbon neutrality. By using smart software to optimize energy use, reduce emissions, and integrate with sustainable systems, these vehicles can play a crucial role in creating a cleaner, greener future

Conclusion

Software-defined vehicles represent a strategic opportunity for automakers to drive innovation, differentiate their offerings, and create value for customers in an increasingly connected and digital world. By embracing a strategic approach that prioritizes investment in talent, partnerships, regulatory compliance, customer-centric design, and agility, automakers can position themselves for success in the era of software-defined vehicles.

FOUR

ARCHITECTURE & OPERATING SYSTEMS

In the ever-evolving landscape of automotive technology, Software Defined Vehicles (SDVs) stand at the forefront, promising a future where software plays a pivotal role in defining the driving experience. But what exactly lies beneath the hood of these innovative machines? Let's take a deep dive into the architecture of SDVs, unraveling the layers that make them tick, and uncovering the key players shaping their development.

1. Chip-Level Architecture:

At the core of every SDV lies a sophisticated network of chips and processors, responsible for executing complex algorithms and handling massive amounts of data in real-time. These chips, often powered by advanced microcontrollers and System-on-Chip (SoC) designs, form the foundation of the vehicle's computing power. Key players in this space include semiconductor giants like NVIDIA, Intel, and Qualcomm, whose cutting-edge technologies enable SDVs to process sensor data, run AI algorithms, and power autonomous driving systems.

2. Operating System (OS) Layer:

Just like your smartphone or computer, SDVs rely on an operating system to manage hardware resources, run applications, and facilitate communication between different components. While traditional automotive OSes were often proprietary and closed, modern SDVs are increasingly adopting open-source platforms like Linux, Android Automotive, and Automotive Grade Linux (AGL). These flexible and customizable OSes provide a foundation for building software-defined features and integrating third-party applications. Key players driving innovation in this space include Google's Android Auto, Microsoft's Automotive Platforms, and Linux Foundation's AGL initiative.

3. Middleware Layer:

Sitting between the OS and application layers, middleware acts as the glue that connects various software components within the SDV ecosystem. This includes communication protocols, data management systems, and software frameworks that enable seamless integration of functionalities like advanced driver-assistance systems (ADAS), infotainment, and vehicle-to-everything (V2X) communication. Companies like Bosch, Continental, and Harman are leading providers of automotive middleware solutions, offering robust platforms and development tools to OEMs and Tier 1 suppliers.

4. Application Layer:

At the top of the SDV architecture sits the application layer, where the magic happens. This encompasses a wide range of software functionalities, from basic vehicle control systems to advanced features like autonomous driving, predictive maintenance, and personalized user experiences. OEMs like Tesla, General Motors, and Volkswagen are driving innovation in this space, developing proprietary software solutions and leveraging partnerships with technology companies to deliver cutting-edge features to consumers.

Key Players Shaping the Future:

Tesla, Inc.: Known for its groundbreaking approach to vehicle software, Tesla leads the pack with its over-the-air (OTA) updates, Autopilot features, and Full Self-Driving (FSD) capabilities.

General Motors (GM): With its Ultium platform and Cruise Automation subsidiary, GM is pushing the boundaries of autonomous driving technology and software-defined vehicle architectures.

Volkswagen Group: Comprising brands like Audi, Porsche, and VW, Volkswagen Group is investing heavily in software-defined vehicle platforms and digital transformation initiatives.

Bosch: As a leading automotive supplier, Bosch plays a crucial role in providing middleware solutions and software development tools to OEMs and Tier 1 suppliers.

NVIDIA Corporation: NVIDIA's powerful computing platforms, such as the DRIVE AGX series, are driving innovation in AI-powered autonomous driving and in-vehicle infotainment systems.

In conclusion, the architecture of Software Defined Vehicles is a complex and dynamic ecosystem, comprised of multiple layers and involving a diverse array of key players. As technology continues to advance and consumer expectations evolve, collaboration between OEMs, suppliers, and technology providers will be essential in shaping the future of automotive innovation.

In the realm of automotive electronics, every Electronic Control Unit (ECU) requires a control program or operating system (OS) to manage a myriad of programs responsible for controlling hardware components and executing various applications specific to each ECU's function. As the complexity of ECUs increases, so does the complexity of the OS required to govern them. Consequently,

Original Equipment Manufacturers (OEMs) must employ multiple OSes to cater to the diverse capabilities and functionalities of ECUs.

For relatively straightforward ECUs, OEMs typically opt for an OS based on the AUTOSAR standard. While AUTOSAR capabilities have advanced, they may not suffice for high-end ECU complexities, such as infotainment systems and most domain-specific ECUs. Notable options for such scenarios include the robust OS offerings from Green Hills and Wind River, renowned for their strong safety and security features.

In contrast, high-end ECUs often rely on QNX or a Linux variant as the OS of choice, with QNX being preferred for applications demanding functional safety. Linux, on the other hand, has emerged as the preferred OS for infotainment systems, surpassing QNX in popularity. Moreover, QNX is gaining traction as the preferred OS for domain-specific ECUs, particularly in the Advanced Driver Assistance Systems (ADAS) and Autonomous Vehicle (AV) domains.

There have been discussions among several OEMs, such as VW and Mercedes-Benz, regarding the development of proprietary car OSes. However, this approach entails significant risks. Developing an OS is a monumental undertaking, with a potential lifespan of 30 to 40 years, necessitating regular updates and continuous technical enhancements. Linux boasts approximately 30 years of development history, while QNX boasts nearly 40 years.

Creating a car OS demands substantial technological expertise, which is often scarce, and entails multiple years of development. General Motors' strategy of adopting Red Hat Linux with functional safety certification presents a more prudent approach to acquiring its own OS for complex ECUs.

In crafting the optimal long-term OS strategy, a best practice entails commencing with the safest OS available for two distinct ECU categories: low complexity and high complexity. This strategy is imperative due to the formidable cybersecurity challenges anticipated in the automotive industry for decades to come, where the choice of OS can significantly impact security.

For low-complexity ECUs, Green Hills stands out with its superior security and safety certifications, including FAA certification for aerospace applications. Meanwhile, for high-end ECUs, QNX boasts higher security and safety certifications compared to Linux variants, owing to its microkernel architecture, which inherently fosters a more secure OS. Furthermore, QNX is likely to lead in developing features aligned with new standards for Autonomous Vehicles (AVs), such as ISO 21448, UL 4600, and IEEE P2851.

In the rapidly evolving landscape of automotive technology, operating systems (OS) play a pivotal role in the realm of software-defined vehicles (SDVs). These sophisticated machines are not merely modes of transportation; they are intricately woven networks of hardware and software, with operating systems serving as the backbone that orchestrates their functionality.

Operating systems for SDVs are tasked with managing an array of complex processes and interactions within the vehicle's ecosystem. From controlling infotainment systems and navigation to regulating safety features such as advanced driver-assistance systems (ADAS), these OSs must seamlessly integrate diverse functionalities to provide a smooth and immersive driving experience.

One of the primary functions of operating systems in SDVs is to facilitate communication between various components and subsystems. Through efficient data exchange protocols, the OS enables sensors, processors, and actuators to interact harmoniously, ensuring real-time responsiveness and adaptability to changing environmental conditions. This interconnectedness allows SDVs to analyze vast amounts of data, ranging from road conditions to vehicle performance metrics, empowering them to make informed decisions and enhance both safety and efficiency on the road.

Moreover, operating systems in SDVs are crucial for managing software updates and maintenance. With the automotive industry witnessing a shift towards over-the-air (OTA) updates, OSs play a vital role in ensuring the seamless delivery and integration of new software features and security patches. This capability not only enhances user convenience by eliminating the need for manual updates but also enables manufacturers to continuously improve and refine the functionality of SDVs throughout their lifecycle.

Security is another critical aspect addressed by operating systems in SDVs. As vehicles become increasingly connected and reliant on external networks, protecting against cyber threats and ensuring data privacy become paramount concerns. Operating systems employ robust security measures, including encryption, authentication, and intrusion detection, to safeguard sensitive information and prevent unauthorized access to vehicle systems.

Furthermore, operating systems for SDVs must be designed with scalability and flexibility in mind. As automotive technology continues to advance at a rapid pace, OSs must be capable of accommodating new hardware configurations, software functionalities, and emerging standards. This adaptability ensures that SDVs remain at the forefront of innovation and can seamlessly integrate with future technologies, such as autonomous driving systems and vehicle-to-everything (V2X) communication.

Linux Based Operating System

In the rapidly evolving landscape of automotive technology, a new era has dawned with the advent of Software-Defined Vehicles (SDVs). At the heart of this transformation lies Linux-based operating systems, providing the foundation upon which the future of mobility is being built. Let's explore how these OSes are shaping the landscape of software-defined vehicles:

Flexibility and Customization: Linux-based operating systems offer unparalleled flexibility and customization, allowing automotive manufacturers to tailor software solutions to meet the unique needs of their vehicles. Whether it's infotainment systems, advanced driver-assistance features, or autonomous driving capabilities, Linux provides a versatile platform that can be adapted to suit a wide range of applications.

Open-Source Collaboration: One of the key advantages of Linux-based OSes is their open-source nature, fostering collaboration and innovation across the automotive industry. Projects like Automotive Grade Linux (AGL) bring together OEMs, suppliers, and developers to create a unified software platform for automotive applications. This collaborative approach ensures interoperability, compatibility, and rapid development of new features and functionalities.

Scalability and Performance: Linux is known for its scalability and performance, making it well-suited for the complex computing requirements of modern vehicles. From low-power embedded systems to high-performance computing platforms, Linux-based OSes can scale seamlessly to meet the demands of software-defined vehicles. This scalability enables the integration of advanced technologies such as artificial intelligence, machine learning, and real-time analytics, paving the way for safer, more efficient transportation systems.

Security and Reliability: Security is paramount in automotive software development, and Linux-based OSes are designed with security in mind. With robust security features, regular updates, and a large community of developers scrutinizing the code, Linux offers a secure and reliable platform for software-defined vehicles. Additionally, real-time operating systems like QNX Neutrino provide deterministic performance and fault tolerance, ensuring the safety and reliability of critical automotive systems.

Cost-Efficiency and Time-to-Market: By leveraging Linux-based OSes, automotive manufacturers can reduce development costs and accelerate time-to-market for new vehicle features and functionalities. The availability of open-source software components and development tools lowers barriers to entry, enabling companies of all sizes to innovate and compete in the rapidly evolving automotive market.

Several Linux-based operating systems are tailored for use in Software Defined Cars, offering flexibility, customization, and scalability to automotive manufacturers. Some notable examples include:

Automotive Grade Linux (AGL): AGL is a collaborative open-source project that provides a Linux-based software platform for automotive applications. It offers a customizable, open-source operating system designed to meet the unique requirements of the automotive industry. AGL is supported by a broad ecosystem of automotive manufacturers, suppliers, and technology companies.

GENIVI Alliance: GENIVI is an open-source software platform that provides a standardized, Linux-based operating system for in-vehicle infotainment (IVI) systems. It aims to accelerate the development and adoption of connected car technologies by providing a common platform for automotive manufacturers and suppliers.

Ubuntu Core: Ubuntu Core is a minimal, containerized version of the Ubuntu Linux distribution designed for embedded and IoT devices. It offers a secure, transactional operating system that can be customized and extended to meet the specific needs of software-defined cars.

QNX Neutrino: While not strictly Linux-based, QNX Neutrino is a Unix-like real-time operating system (RTOS) commonly used in automotive infotainment and telematics systems. It offers a scalable, reliable platform with support for advanced graphics, multimedia, and connectivity features.

Wind River Linux: Wind River Linux is a commercial Linux distribution tailored for embedded and IoT applications, including automotive systems. It provides a customizable, secure operating system platform with long-term support and maintenance services for automotive manufacturers and suppliers.

Red Hat OpenShift: Red Hat's Kubernetes-based container platform, OpenShift, is increasingly being used in automotive software development. OpenShift enables developers to build, deploy, and manage containerized applications across hybrid cloud environments, making it well-suited for developing and deploying software-defined car applications that require scalability, portability, and agility.

These Linux-based operating systems offer a solid foundation for building software-defined cars, providing support for a wide range of automotive applications, from infotainment and navigation to advanced driver assistance systems (ADAS) and autonomous driving features.

In conclusion, Linux-based operating systems are playing a pivotal role in shaping the future of software-defined vehicles. With their flexibility, openness, scalability, security, and cost-efficiency, Linux-based OSes empower automotive manufacturers to create the next generation of intelligent, connected, and autonomous vehicles. As we continue to push the boundaries of innovation, Linux will remain at the forefront of driving positive change in the automotive industry.

Non Linux based operating Systems

In the dynamic realm of Software Defined Vehicles (SDVs), operating systems serve as the backbone of automotive software solutions, dictating functionality, performance, and user experience. While Linux-based OSes dominate this space, several non-Linux alternatives are also pivotal in automotive innovation. Let's delve into these operating systems and explore how they shape the future of SDVs, highlighting key OEMs and Tier 1 suppliers leveraging them.

1. QNX Neutrino: Powering Safety-Critical Systems

Example OEMs and Tier 1 Suppliers:
OEM: Ford Motor Company
Tier 1 Supplier: Harman International

Renowned for its real-time capabilities, QNX Neutrino by BlackBerry is a go-to choice for safety-critical systems in SDVs. With OEMs like Ford Motor Company incorporating QNX Neutrino in their SYNC infotainment systems, and Tier 1 suppliers like Harman International integrating it into advanced telematics solutions, QNX Neutrino ensures reliability and determinism in critical automotive applications.

2. Microsoft Automotive: Driving Infotainment Innovation

Example OEMs and Tier 1 Suppliers:
OEM: Volvo Cars
Tier 1 Supplier: Panasonic Automotive Systems

Microsoft Automotive solutions, including Windows Automotive Embedded, are synonymous with rich infotainment experiences.

OEMs like Volvo Cars leverage Microsoft's technology to power their Sensus infotainment system, while Tier 1 suppliers like Panasonic Automotive Systems integrate it into their in-car entertainment solutions, showcasing Microsoft's role in driving infotainment innovation in SDVs.

3. RTOS-based Systems: Ensuring Real-Time Responsiveness

Example OEMs and Tier 1 Suppliers:
OEM: BMW Group
Tier 1 Supplier: Bosch Automotive

RTOS-based systems such as FreeRTOS and VxWorks are instrumental in ensuring real-time responsiveness in automotive control systems. BMW Group employs VxWorks in its advanced driver assistance systems (ADAS), while Tier 1 suppliers like Bosch Automotive integrate FreeRTOS into their electronic control units (ECUs), demonstrating the critical role of RTOS in enabling precise control and coordination of vehicle functions.

4. Custom Proprietary Operating Systems: Tailored for Unique Needs

Example OEMs and Tier 1 Suppliers:
OEM: Tesla, Inc.
Tier 1 Supplier: Continental AG

Custom proprietary operating systems cater to unique requirements, exemplified by OEMs like Tesla, Inc., which develops its software stack, including a custom OS, to power its electric vehicles. Similarly, Tier 1 suppliers like Continental AG develop bespoke operating systems for their advanced driver assistance systems (ADAS) and vehicle connectivity solutions, showcasing the flexibility and customization offered by proprietary OS solutions.

5. Android Automotive: Bringing Familiarity and Connectivity

Example OEMs and Tier 1 Suppliers:
OEM: Volvo Cars
Tier 1 Supplier: Panasonic Automotive Systems

Android Automotive, optimized for in-vehicle infotainment, enhances connectivity and user experience. OEMs like Volvo Cars integrate Android Automotive into their infotainment systems, offering features such as Google Assistant integration and app ecosystem access. Panasonic Automotive Systems, a Tier 1 supplier, leverages Android Automotive in its connected car solutions, exemplifying the platform's role in driving innovation in SDVs.

In conclusion, while Linux-based operating systems remain prevalent in SDVs, non-Linux alternatives are instrumental in specific automotive applications. From safety-critical systems to infotainment innovation, these operating systems, embraced by key OEMs and Tier 1 suppliers, contribute to the diverse tapestry of automotive software solutions, shaping the future of SDVs in exciting and impactful ways.

Devops Strategies

In recent years, the automotive industry has witnessed a paradigm shift with the emergence of Software Defined Vehicles (SDVs). These vehicles, equipped with advanced software and connectivity features, are redefining transportation as we know it. However, with this evolution comes the challenge of efficiently managing and updating the complex software systems that power SDVs. Enter DevOps – a set of practices that blend software development (Dev) with IT operations (Ops) to streamline software delivery and enhance collaboration. In this article, we explore how DevOps is transforming SDVs and highlight key vendors, OEMs, and Tier 1 companies leading the charge in this revolution.

SDVs represent a new era in automotive technology, where software plays a central role in defining the vehicle's functionality, performance, and user experience. From advanced driver-assistance systems (ADAS) to infotainment features and over-the-air (OTA) updates, software is driving innovation and differentiation in the automotive industry. However, the traditional methods of software development and deployment are often incompatible with the dynamic and interconnected nature of SDVs, necessitating a new approach – DevOps.

DevOps: Enabling Agility and Innovation

DevOps principles emphasize automation, collaboration, and continuous improvement, making it an ideal fit for the fast-paced world of SDVs. By adopting DevOps practices, automotive companies can achieve the following benefits:

Accelerated Time-to-Market: DevOps enables rapid development, testing, and deployment of software updates, allowing automotive companies to deliver new features and enhancements to market faster.

Improved Quality and Reliability: Continuous integration and continuous delivery (CI/CD) pipelines automate testing and validation processes, reducing the risk of software errors and improving the overall reliability of SDVs.

Enhanced Collaboration: DevOps encourages cross-functional collaboration between development, operations, and other stakeholders, fostering a culture of shared responsibility and innovation.

Key Vendors and Companies Embracing DevOps for SDVs

1. Tesla, Inc.

Tesla has been at the forefront of integrating software into vehicles, with features like Autopilot and Full Self-Driving (FSD).

The company utilizes DevOps practices to deploy OTA updates seamlessly, delivering new features and improvements to its fleet of vehicles.

2. General Motors (GM)

GM has invested heavily in software-defined vehicle platforms, such as its Ultium platform for electric vehicles.

The company leverages DevOps to accelerate software development and enable OTA updates for critical vehicle systems.

3. Volkswagen Group

Volkswagen has embraced DevOps to drive digital transformation across its brands, including Audi, Porsche, and Volkswagen.

DevOps practices enable Volkswagen to rapidly iterate on software features and deliver personalized experiences to customers.

4. Bosch

As a leading automotive supplier, Bosch plays a crucial role in developing software-defined solutions for vehicle manufacturers.

The company utilizes DevOps to streamline software development and ensure compatibility with a wide range of vehicle platforms.

5. NVIDIA Corporation

NVIDIA's platforms, such as DRIVE, power advanced AI and computing capabilities in SDVs.

DevOps practices enable NVIDIA to deliver software updates and optimizations for its automotive platforms, enhancing performance and safety.

Conclusion

As Software Defined Vehicles continue to reshape the automotive landscape, DevOps emerges as a critical enabler of innovation, agility, and collaboration. By embracing DevOps principles and practices, automotive companies can unlock the full potential of SDVs, delivering superior experiences to customers and driving the industry forward. With key vendors, OEMs, and Tier 1 companies leading the charge, the future of transportation promises to be more connected, intelligent, and exciting than ever before.

In conclusion, operating systems are the unsung heroes powering the revolution in software-defined vehicles. From orchestrating complex interactions to ensuring security and scalability, OSs play a vital role in shaping the capabilities and performance of SDVs. As the automotive industry continues to embrace digital transformation, operating systems will remain indispensable components, driving innovation and redefining the future of mobility.

FIVE

OPEN SOURCE DEVELOPMENT: THE BLOCKBUSTER FOR SOFTWARE-DEFINED VEHICLES?

In the ever-evolving landscape of technology and innovation, every industry dreams of finding its "blockbuster" breakthrough—a game-changer that revolutionizes the way things are done. In the world of movies, a blockbuster captivates audiences, breaking box office records and leaving a lasting impact on the cinematic landscape.

Similarly, industries like pharmaceuticals and automotive are constantly on the lookout for that disruptive "blockbuster drug" or innovation that can change the game. Could open source development be the blockbuster that the automotive industry needs to propel it into the future of software-defined vehicles?

Imagine this: Just like a blockbuster movie captivates audiences with its unexpected twists and turns, open source development has the potential to disrupt the automotive industry in ways we can't yet fully comprehend. It's the unexpected plot twist that shakes up the status quo and challenges traditional norms. But like any good movie, there are both advantages and disadvantages to this approach.

Hollywood thrives on blockbusters – movies that shatter box office records and redefine genres. In the world of software-defined vehicles (SDVs), the industry desperately seeks its own blockbuster: a disruptive force that rewrites the script on car development. Could open-source development be the silver screen sensation that propels SDVs into the future?

The Current Scene: A Closed Shop with Limited Showtimes

Imagine a movie studio where every scene is shot behind closed doors, and only a select few get to witness the final product. That's the current state of automotive software development. Original Equipment Manufacturers (OEMs) and Tier 1 suppliers operate in a siloed environment, each clinging to proprietary code like a closely guarded secret. This stifles innovation and slows down development, leaving the audience (consumers) yearning for a more dynamic and feature-rich experience.

Now, let's talk about the key players in this blockbuster drama: the OEMs (Original Equipment Manufacturers) and Tier 1 suppliers.

OEMs, like leading actors in a movie, play a central role in shaping the direction of the storyline. They are responsible for defining the overall vision and strategy for their vehicles, including the software architecture and features. For OEMs, open source development represents both an opportunity and a challenge. On one hand, it offers the potential for greater innovation and differentiation. On the other hand, it requires navigating complex partnerships and managing the risks associated with open source licensing and intellectual property.

Tier 1 suppliers, like supporting cast members, provide essential components and systems that contribute to the overall performance and functionality of the vehicle. They are often tasked with implementing and integrating software solutions from various sources, including open source projects. For Tier 1 suppliers, open source development presents an opportunity to collaborate with OEMs and other industry players to deliver cutting-edge technologies and solutions.

But not everyone in this blockbuster saga is rooting for open source development. Some stakeholders may be skeptical or resistant to change, fearing the potential disruption it could bring to established business models and practices. Traditionalists may prefer the comfort and familiarity of proprietary software solutions, while others may have legitimate concerns about security, reliability, and intellectual property rights.

Open Source: The Hero Enters Stage Left

Open-source software development bursts onto the scene as a potential blockbuster. It's a radical shift, akin to throwing open the studio doors and inviting a global audience of developers to contribute. This collaborative approach promises:

Faster Innovation: With a legion of minds working together, code development goes into hyperdrive, churning out features and functionalities at breakneck speed.

Reduced Costs: Sharing the development burden lightens the load for everyone – a welcome change for cash-strapped car companies.

Enhanced Security: A community of watchful eyes scrutinizes the code, identifying and patching vulnerabilities quicker than a lone team ever could.

The Plot Thickens: Not Everyone Loves a Sequel

However, the open-source revolution isn't without its plot twists. Here are the villains lurking in the shadows:

Safety Concerns: The automotive industry is built on a foundation of safety. Can open-source development, with its inherent fluidity, guarantee the rigorous testing and certification required for critical software?

Intellectual Property (IP) Woes: OEMs, wary of diluting their competitive edge, hesitate to share code that differentiates their vehicles. Will they be willing to give up their secret sauce for the sake of collaboration?

Who's In, Who's Out? Picking Sides in the Blockbuster Brawl

So, who will be the heroes and villains of this automotive drama? Here's a breakdown:

In Favor: Tech-savvy startups and forward-thinking OEMs see the potential for rapid innovation and cost reduction.

Against: Traditional Tier 1s, heavily invested in existing closed ecosystems, might resist the open-source wave.

The Climax: A Collaborative Future, But With Nuances

The most likely outcome? A blockbuster with a twist. Open-source development won't be a silver bullet, but it will be a powerful tool in the SDV arsenal. We can expect a hybrid approach, with a mix of open-source platforms and proprietary code used strategically.

CEOs, Take Note: Your Director's Chair Awaits

The future of SDVs rests in the hands of visionary CEOs. Here's your action plan:

Embrace Collaboration: Foster a culture of open innovation, partnering with Tier 1s and fostering an internal culture that thrives on collaboration.

Strategize for Open Source: Identify core functionalities that can benefit from open-source development, while safeguarding critical IP.

Invest in Security: Develop robust security protocols to ensure the integrity and safety of open-source code within your SDV framework.

Open-source development may not be a guaranteed blockbuster, but it offers a compelling plotline for the future of SDVs. By embracing collaboration and playing their roles strategically, CEOs can rewrite the script, ensure a successful run at the box office, and ultimately deliver the hit car of tomorrow.

SIX

REGULATIONS & CYBER SECURITY CHALLENGES

Regulations governing Software Defined Vehicles (SDVs) are critical to ensure safety, cybersecurity, and compliance with industry standards. As SDVs increasingly rely on software for their operation, regulatory bodies worldwide are developing and updating regulations to address the unique challenges posed by these advanced vehicles. Here's an overview of key automotive regulations applicable to SDVs:

1. Functional Safety Standards

ISO 26262: This international standard specifies functional safety requirements for automotive systems, including SDVs. It outlines processes for hazard analysis, risk assessment, and functional safety measures throughout the vehicle's lifecycle to mitigate safety risks associated with software and electronic systems.

2. Cybersecurity Regulations

UNECE WP.29: The United Nations Economic Commission for Europe (UNECE) WP.29 regulations include cybersecurity provisions for vehicle type approval. These regulations address cybersecurity threats and require manufacturers to implement measures to protect SDVs from cyberattacks and unauthorized access to vehicle systems.

3. Software Development Guidelines

MISRA C: The Motor Industry Software Reliability Association (MISRA) C guidelines provide best practices for the development of safe and reliable software in automotive systems. Adhering to these guidelines helps ensure the integrity and quality of software used in SDVs.

4. Data Protection and Privacy Regulations

GDPR: The General Data Protection Regulation (GDPR) in the European Union (EU) governs the collection, processing, and storage of personal data, including data generated by SDVs. Compliance with GDPR is essential to protect consumer privacy and ensure responsible handling of vehicle data.

5. OTA Update Guidelines

OTA Update Guidelines: Regulatory bodies are developing guidelines for over-the-air (OTA) software updates in vehicles. These guidelines address aspects such as update verification, data security, and transparency to ensure the safe and reliable delivery of software updates to SDVs.

6. Autonomous Vehicle Regulations

NHTSA Regulations (US): The National Highway Traffic Safety Administration (NHTSA) in the United States is developing regulations specific to autonomous vehicles, including SDVs. These regulations cover various aspects of autonomous driving systems, testing procedures, and safety requirements to ensure the safe deployment of SDVs on public roads.

7. Liability and Insurance Regulations

Product Liability Laws: Liability frameworks are evolving to address questions of responsibility in the event of accidents or malfunctions involving SDVs. Regulatory bodies and lawmakers are considering changes to product liability laws to determine liability in cases where software or autonomous driving systems are involved.

Regulatory compliance is essential for the safe and responsible deployment of Software Defined Vehicles. By adhering to functional safety standards, cybersecurity regulations, software development guidelines, and data protection laws, manufacturers can ensure the integrity, safety, and security of SDVs, fostering trust among consumers and regulatory authorities alike.'

As Software driven vehicles (SDVs) become increasingly integrated into our transportation systems, ensuring robust cybersecurity measures is paramount. These vehicles rely on complex systems of sensors, software, and networks, making them vulnerable to cyber threats. To address these challenges, a comprehensive cybersecurity framework is essential, incorporating rules, laws, innovative products from leading vendors, and rigorous testing methodologies.

Rules and Laws: Governments worldwide are enacting regulations to address cybersecurity concerns in SDVs. These rules mandate stringent security standards and data protection measures. For instance, the General Data Protection Regulation (GDPR) in Europe and the California Consumer Privacy Act (CCPA) in the United States impose strict guidelines on data handling and breach notifications, which impact SDV manufacturers and service providers.

Key Products and Vendors:

Argus Cyber Security: Argus provides comprehensive cybersecurity solutions tailored for connected vehicles. Their products focus on intrusion detection, prevention, and secure over-the-air updates, ensuring the integrity and safety of SDV systems.

GuardKnox: GuardKnox offers secure automotive solutions, including hardware and software platforms designed to protect vehicle networks from cyber threats. Their products incorporate multi-layered defenses and real-time monitoring to safeguard critical systems.

Palo Alto Networks: Palo Alto Networks delivers advanced firewall and threat prevention solutions essential for securing SDV networks. Their Next-Generation Firewall (NGFW) technology offers granular control and visibility, enabling proactive threat mitigation.

Red Hat: Red Hat provides open-source software solutions crucial for SDV cybersecurity. Their Linux-based operating systems offer robust security features and support for containerization, enhancing the resilience of SDV software ecosystems.

Trillium Secure: Trillium specializes in cybersecurity platforms for connected vehicles, offering solutions for threat detection, authentication, and secure communication. Their products mitigate risks associated with vehicle-to-everything (V2X) connectivity and data exchange.

Benefits and Risks: The adoption of cybersecurity products in SDVs offers numerous benefits, including:

Protection against cyber threats: These products help prevent unauthorized access, data breaches, and malicious attacks, ensuring the safety and reliability of SDV systems.

Compliance with regulations: By implementing cybersecurity solutions, SDV manufacturers can comply with industry standards and regulatory requirements, avoiding potential legal liabilities.

Enhanced trust and reputation: Robust cybersecurity measures inspire confidence among consumers, stakeholders, and regulatory authorities, fostering trust in SDV technologies and services.

However, the risks associated with cybersecurity vulnerabilities in SDVs are significant:

Safety concerns: Cyberattacks on SDVs can compromise vehicle functionality, leading to accidents, injuries, or loss of life. Malicious actors could manipulate sensor data, disrupt control systems, or remotely hijack vehicles, posing serious safety risks.

Privacy implications: Breaches of sensitive data stored in SDVs, such as location information, driver profiles, and vehicle telemetry, raise privacy concerns and may result in identity theft, surveillance, or unauthorized surveillance.

Cybersecurity Development Process: The development of cybersecurity solutions for SDVs typically involves several stages, including:

Threat Analysis and Risk Assessment (TARA): This phase identifies potential cyber threats, vulnerabilities, and attack vectors in SDV systems. TARA helps prioritize security measures and allocate resources effectively to mitigate risks.

Penetration Testing: Penetration testing involves simulating cyber attacks to assess the resilience of SDV systems against real-world threats. Ethical hackers attempt to exploit vulnerabilities in software, networks, and hardware components, providing valuable insights for improving security controls and defenses.

Key OEMs and Tier Suppliers: Original Equipment Manufacturers (OEMs) and Tier suppliers play a crucial role in ensuring strong cybersecurity across the automotive supply chain. Companies such as Toyota, BMW, Volkswagen, and Bosch prioritize cybersecurity in their SDV development processes, collaborating with technology partners, researchers, and regulatory agencies to address emerging threats and vulnerabilities.

Real case scenario analysis

Upstream conducted a thorough analysis of 295 publicly disclosed cyberattacks in 2023, revealing a slight increase from the previous year. Within this tally, a significant 64% were attributed to malicious Black Hat hackers. However, more crucial than the sheer number of breaches, according to Sarid-Hausirer, is the modus operandi employed and the specific targets of these cybersecurity intruders.

Illustratively, a staggering 85% of these breaches were executed remotely – with no physical contact with the compromised vehicle, marking a notable surge from the 70% recorded in 2022. This indicates that attacks can originate from any corner of the globe.

What's even more alarming is the potential scale of impact on vehicles vulnerable to these cyber threats. Approximately 49% of the breaches targeted a high (in the thousands) or massive (in the millions) number of vehicles on the road, more than doubling the 23% penetration rate observed in 2022. When focusing solely on the 135 Black Hat attacks, the potential for high/massive impacts surged to 67%.

A concerning revelation is that 13% of these Black Hat attacks utilized application programming interfaces (APIs) – the conduits enabling software-to-software or cloud communication – a marginal increase from 2022. However, the expanding scope of vehicles interconnected via APIs due to the rising prevalence of software-defined vehicles indicates a looming threat, as such attacks hold the potential to affect a substantial number of vehicles in the near future.

The majority of the Black Hat attacks in 2023 targeted multiple Original Equipment Manufacturers (OEMs), with 59% focusing on various global or regional automakers concurrently.

Particularly worrisome for consumers is that 12% of the Black Hat attacks specifically aimed to obtain personal information accessible through a vehicle's applications, putting sensitive data like credit card numbers and financial details at risk.

Furthermore, malicious actors are increasingly leveraging Generative Artificial Intelligence (AI) to expedite and broaden the impact of cyberattacks. This innovation streamlines the attack process and enhances attackers' understanding of software intricacies, thereby reducing barriers to entry for potential assailants.

In one scenario, Upstream estimates that an OEM could incur costs of up to $50 million if one of its battery-electric vehicles were hacked through its battery-management or charging software. This projection encompasses expenses associated with vehicle and battery recalls, repairs, as well as legal and regulatory compliance, including potential class-action lawsuits.

To counter these escalating threats, Upstream advocates for proactive measures, including safeguarding data not just within vehicles but also in the cloud. Additionally, continuous monitoring of the Dark Web, where malicious actors often congregate and share information, is recommended. Embracing a "shift left" approach that embeds cybersecurity protection into the OEM's product development chain from the outset, extending into the R&D department, is crucial for bolstering defense mechanisms.

Conclusion

Cybersecurity is integral to the advancement and adoption of self-driving vehicles, safeguarding passengers, pedestrians, and infrastructure from potential cyber threats. By adhering to regulatory requirements, leveraging innovative products from leading vendors, and adopting robust testing methodologies, the automotive industry can build resilient and secure SDV ecosystems for the future.

SEVEN

IMPACT OF GENERATIVE AI

There are two set of ecosystems emerging in Global markets and Chinese markets. GENAI, poised to revolutionize Software Defined Vehicles (SDVs), is set to unleash a wave of transformative capabilities in the automotive industry. With its advanced AI-driven solutions, GENAI promises to elevate the driving experience, enhance safety, and optimize vehicle performance. Here are five compelling use cases where GENAI, in collaboration with leading technology vendors, is set to redefine the future of SDVs,

Predictive Maintenance with NVIDIA/HUAWEI: By harnessing their powerful AI algorithms, GENAI enables predictive maintenance in SDVs. Through real-time analysis of vehicle data, including sensor readings and performance metrics, GENAI predicts potential mechanical issues before they occur, allowing for proactive maintenance and minimizing downtime.

Enhanced Driver Assistance Systems (ADAS) with Intel/Baidu : Leveraging their cutting-edge processing power, GENAI enhances ADAS capabilities in SDVs. With AI-powered algorithms, GENAI enables intelligent decision-making for features like adaptive cruise control, lane-keeping assistance, and collision avoidance, significantly improving driver safety.

Personalized In-Car Experience with IBM Watson/Tencent AI: Partnering with them, GENAI offers a personalized in-car experience for passengers. Using natural language processing and machine learning, GENAI understands and responds to passenger preferences, adjusting climate control, entertainment options, and even suggesting nearby attractions based on individual preferences.

Real-Time Traffic Optimization with Google Cloud/Alibaba Cloud: Collaborating with them, GENAI optimizes real-time traffic routing in SDVs. By analyzing traffic data, weather conditions, and road incidents, GENAI dynamically adjusts navigation routes to minimize travel time, reduce congestion, and enhance overall efficiency.

Cybersecurity Solutions with Red Hat/Xiaomi Security: Teaming up with them, GENAI strengthens cybersecurity measures in SDVs. Through robust encryption protocols, intrusion detection systems, and continuous monitoring, GENAI ensures the integrity and security of vehicle software, protecting against cyber threats and safeguarding sensitive data.

In essence, GENAI, in synergy with leading technology vendors, is poised to revolutionize SDVs by delivering unparalleled safety, efficiency, and personalized experiences on the road. As the automotive industry embraces AI-driven innovations, the future of driving has never looked more promising.

Generative AI (GenAI) has the potential to significantly impact Software-Defined Vehicles (SDVs) in several ways, potentially revolutionizing the automotive industry. Here's a breakdown of the potential effects:

1. Design and Development:

Automated Design: GenAI can be used to generate and iterate on car designs much faster than traditional methods. This could lead to a wider variety of innovative and user-friendly vehicle designs tailored to specific needs.

Personalized Interiors: GenAI could personalize car interiors based on user preferences, generating layouts and recommending features that optimize comfort and functionality for each driver.

Predictive Maintenance: GenAI can analyze sensor data from the vehicle to predict potential maintenance issues and even generate personalized repair plans.

2. Software Optimization:

Self-Optimizing Software: GenAI can analyze driving patterns and traffic conditions to optimize software performance for efficiency, safety, and even comfort. This could involve adjusting powertrain management systems, suspension settings, and even climate control.

Bug Detection and Prevention: GenAI can analyze vast amounts of code used in SDVs to identify potential bugs and vulnerabilities before they cause problems. This could lead to more reliable and secure software for vehicles.

3. Advanced Driver-Assistance Systems (ADAS):

Enhanced Situational Awareness: GenAI can analyze data from various sensors (cameras, LiDAR, radar) in real-time, providing a more comprehensive understanding of the surroundings and improving the performance of ADAS features.

Personalized Driving Assistance: GenAI could personalize ADAS functionalities based on driver behavior and preferences. For instance, adjusting lane departure warnings or adaptive cruise control sensitivity.

4. User Experience:

Natural Language Interaction: GenAI could power more natural and intuitive voice assistants within vehicles, enabling seamless interaction with the car's various systems.

Personalized In-Car Entertainment: GenAI could recommend music, movies, or audiobooks based on user preferences and driving conditions, tailoring the in-car entertainment experience.

Challenges and Considerations:

Safety and Reliability: Safety is paramount in the automotive industry. Stringent testing and validation will be crucial to ensure GenAI-powered features are safe and reliable.

Data Privacy: The use of GenAI will involve collecting and analyzing vast amounts of user data. Robust data security and privacy measures will be essential.

Ethical Considerations: GenAI algorithms can perpetuate biases. Careful design and training of AI models will be necessary to avoid biased decision-making in areas like ADAS functionalities.

Overall, GenAI has the potential to be a game-changer for SDVs. However, responsible development and implementation are key to ensuring safety, reliability, and ethical considerations are addressed. By harnessing the power of GenAI, car manufacturers can create smarter, more personalized, and ultimately safer driving experiences for the future.

EIGHT

CHIP TO CLOUD STRATEGIES

In the realm of automotive innovation, the convergence of technology and mobility has given rise to a new paradigm: software-defined vehicles. These vehicles, equipped with an array of sensors, processors, and connectivity solutions, are transforming the way we perceive transportation. At the heart of this transformation lies the chip-to-cloud strategy, a comprehensive approach that leverages the power of integrated circuits (chips) and cloud computing to enhance the capabilities of modern vehicles.

Understanding Software-Defined Vehicles

Software-defined vehicles represent a departure from traditional automotive design, where hardware components played a dominant role. In contrast, these vehicles rely heavily on software to control and optimize various functions, from propulsion and braking to infotainment and autonomous driving. This shift towards software-defined architecture offers several benefits, including flexibility, scalability, and the ability to adapt to evolving consumer demands.

The Role of Chips in Automotive Innovation

Central to the chip-to-cloud strategy is the integration of advanced semiconductor technology within vehicles. These chips serve as the computational backbone, enabling real-time data processing, sensor fusion, and decision-making capabilities. Key components include microcontrollers, graphics processing units (GPUs), and application-specific integrated circuits (ASICs), each optimized for specific tasks such as image recognition, sensor data analysis, and predictive maintenance.

Leveraging Cloud Computing for Enhanced Capabilities

While onboard chips provide local processing power, the true potential of software-defined vehicles lies in their ability to harness the power of the cloud. By establishing seamless connectivity to cloud-based resources, vehicles can access vast amounts of data, perform complex computations, and benefit from advanced machine learning algorithms. This enables features such as over-the-air (OTA) updates, predictive analytics, and personalized services tailored to individual driver preferences.

Advantages of the Chip-to-Cloud Strategy

Real-Time Insights: By processing data locally and leveraging cloud resources, software-defined vehicles can gain real-time insights into their surroundings, enabling faster response times and improved safety.

Scalability: The chip-to-cloud architecture allows for scalable deployment of new features and services, ensuring that vehicles remain relevant and up-to-date throughout their lifecycle.

Cost-Efficiency: By offloading certain computational tasks to the cloud, automakers can reduce the need for expensive onboard hardware, leading to cost savings and improved affordability for consumers.

Addressing Challenges and Concerns

While the chip-to-cloud strategy offers numerous benefits, it also presents challenges that must be addressed. These include concerns regarding data privacy and security, as well as the need for robust connectivity infrastructure to ensure reliable communication between vehicles and the cloud. Additionally, there may be regulatory hurdles to overcome, particularly regarding the certification and validation of software-driven automotive systems.

In automotive applications, a variety of chips are used to power different systems and functionalities within vehicles. Some common types of chips include:

Microcontrollers (MCUs): These are the workhorses of automotive electronics, responsible for controlling various functions such as engine management, braking systems, and powertrain control. MCUs are highly optimized for real-time processing and often feature integrated peripherals such as analog-to-digital converters (ADCs) and communication interfaces like CAN (Controller Area Network) and LIN (Local Interconnect Network).

Graphics Processing Units (GPUs): GPUs are essential for powering advanced infotainment systems and in-vehicle displays, providing high-quality graphics rendering for navigation, entertainment, and driver assistance features.

Digital Signal Processors (DSPs): DSPs are specialized chips designed for processing digital signals in real-time. In automotive applications, DSPs are used for tasks such as audio processing, noise cancellation, and sensor data filtering.

Field-Programmable Gate Arrays (FPGAs): FPGAs offer flexibility and programmability, making them ideal for prototyping and customization in automotive electronics. They are commonly used for tasks such as sensor interfacing, data preprocessing, and rapid system prototyping.

Application-Specific Integrated Circuits (ASICs): ASICs are custom-designed chips optimized for specific automotive applications. They are commonly used in safety-critical systems such as airbag control modules, anti-lock braking systems (ABS), and advanced driver assistance systems (ADAS).

Memory Chips: Various types of memory chips, including Flash memory and DRAM, are used in automotive electronics for storing program code, data, and configuration settings. These chips are essential for supporting OTA updates, vehicle diagnostics, and multimedia storage.

Sensors and Sensor Interface Chips: Automotive sensors, such as accelerometers, gyroscopes, and radar modules, rely on dedicated sensor interface chips to convert analog signals into digital data that can be processed by the vehicle's electronic control units (ECUs).

Power Management ICs: These chips regulate voltage levels and manage power distribution within the vehicle's electrical system, ensuring optimal performance and efficiency while minimizing energy consumption and heat generation.

These are just a few examples of the many types of chips used in automotive electronics. As vehicles become increasingly connected, electrified, and automated, the demand for specialized semiconductor solutions tailored to automotive applications will continue to grow.

Who are the key players in the Chip strategy?

Microcontrollers (MCUs):

Key Players:

NXP Semiconductors: Renowned for its high-performance MCUs tailored for automotive applications, NXP offers a wide range of solutions designed to meet stringent safety, reliability, and efficiency requirements.

Infineon Technologies: Infineon is a leading provider of automotive-grade MCUs, offering innovative solutions for powertrain control, chassis systems, and advanced driver assistance systems (ADAS).

STMicroelectronics: STMicroelectronics is a key player in the automotive MCU market, offering a comprehensive portfolio of products optimized for automotive applications such as engine management, body control, and infotainment systems.

Graphics Processing Units (GPUs):

Key Players:

NVIDIA: NVIDIA is a dominant force in the GPU market, providing cutting-edge solutions for automotive infotainment, navigation, and driver assistance systems. Its powerful GPUs enable immersive graphics rendering and real-time processing for enhanced user experiences.

Qualcomm: Qualcomm's Snapdragon Automotive platform integrates powerful GPUs with advanced connectivity and AI capabilities, enabling automotive OEMs to deliver next-generation infotainment and telematics solutions.

Renesas Electronics: Renesas offers a range of automotive-grade GPUs optimized for display systems, instrument clusters, and in-vehicle entertainment. Its solutions provide high-performance graphics rendering and support for multiple displays.

Digital Signal Processors (DSPs):

Key Players:

Texas Instruments (TI): TI is a leading supplier of automotive DSPs, offering solutions for audio processing, voice recognition, and sensor data processing. Its DSPs are widely used in automotive infotainment, communication, and driver assistance systems.

Analog Devices: Analog Devices provides high-performance DSPs tailored for automotive applications, enabling advanced audio processing, noise cancellation, and real-time signal processing in vehicles.

NXP Semiconductors: NXP offers a range of automotive-grade DSPs optimized for audio processing, speech recognition, and sensor fusion. Its DSP solutions are integral to delivering immersive in-vehicle audio experiences and enhancing safety features.

Field-Programmable Gate Arrays (FPGAs):

Key Players:

Xilinx: Xilinx is a prominent player in the FPGA market, providing automotive-grade solutions for rapid prototyping, sensor interfacing, and data preprocessing. Its FPGAs enable flexibility and customization in automotive electronics design.

Intel Corporation: Intel's FPGA solutions, including its Intel® Agilex™ FPGAs, are utilized in automotive applications such as advanced driver assistance systems, autonomous driving, and in-vehicle networking. Intel's FPGAs offer high-performance computing and reconfigurability for automotive innovation.

Microchip Technology: Microchip offers a range of automotive-grade FPGAs designed for reliability, performance, and low power consumption. Its solutions are used in automotive control systems, sensor interfaces, and connectivity applications.

Application-Specific Integrated Circuits (ASICs):

Key Players:

STMicroelectronics: STMicroelectronics is a leading provider of automotive ASICs, offering custom-designed solutions for safety-critical applications such as airbag control modules, anti-lock braking systems (ABS), and advanced driver assistance systems (ADAS).

Renesas Electronics: Renesas specializes in automotive ASICs optimized for performance, reliability, and safety. Its solutions address a wide range of automotive applications, including powertrain control, body electronics, and vehicle networking.

ON Semiconductor: ON Semiconductor delivers automotive-grade ASICs for various applications, including motor control, lighting systems, and battery management. Its custom-designed ASICs enable efficient and reliable operation in harsh automotive environments.

Memory Chips:

Key Players:

Samsung Electronics: Samsung is a major player in the memory chip market, offering a comprehensive portfolio of Flash memory and DRAM solutions for automotive applications. Its memory chips provide high-speed data storage and reliable performance in automotive electronics.

SK Hynix: SK Hynix supplies automotive-grade memory chips optimized for reliability, endurance, and temperature tolerance. Its Flash memory and DRAM solutions support critical functions such as OTA updates, vehicle diagnostics, and multimedia storage.

Micron Technology: Micron provides memory solutions tailored for automotive requirements, including Flash memory, LPDDR4/5 DRAM, and NOR Flash. Its memory chips offer high-density storage, low power consumption, and robust performance for automotive applications.

Sensors and Sensor Interface Chips:

Key Players:

Bosch Sensortec: Bosch Sensortec is a leading supplier of automotive sensors and sensor interface chips, offering solutions for accelerometers, gyroscopes, pressure sensors, and environmental sensors. Its products enable precise measurement and reliable data acquisition for vehicle control and safety systems.

STMicroelectronics: STMicroelectronics provides a wide range of automotive sensors and sensor interface chips, including MEMS sensors, magnetic sensors, and interface ICs. Its solutions support various automotive applications, such as navigation, motion detection, and tire pressure monitoring.

Analog Devices: Analog Devices offers automotive-grade sensor interface chips optimized for signal conditioning, amplification, and data conversion. Its interface ICs enable accurate measurement and reliable communication with automotive sensors, enhancing overall system performance and reliability.

Power Management ICs:

Key Players:

Infineon Technologies: Infineon is a leading provider of automotive power management ICs, offering solutions for voltage regulation, power distribution, and energy efficiency. Its products ensure stable power supply and optimal energy management in automotive systems.

Texas Instruments (TI): TI delivers automotive-grade power management ICs for various applications, including battery management, motor control, and LED lighting. Its solutions provide efficient power conversion, protection, and monitoring for automotive electronics.

ON Semiconductor: ON Semiconductor offers a comprehensive portfolio of automotive power management ICs, including voltage regulators, DC-DC converters, and load switches. Its products enable efficient power distribution, thermal management, and protection in automotive systems.

Chinese & Taiwanese companies' ecosystem in Chip Strategy

Microcontrollers (MCUs):

Key Players:

Horizon Robotics (China) : Its a home grown company and will be a major player in times to come.

NXP Semiconductors (China): NXP has a significant presence in China, providing high-performance MCUs tailored for automotive applications. Its solutions are widely used in Chinese vehicles for engine management, braking systems, and powertrain control.

Freescale Semiconductor (China): Freescale, now a part of NXP, offers a range of automotive-grade MCUs designed for Chinese automakers. Its MCUs are optimized for real-time processing and feature integrated peripherals for enhanced functionality.

MediaTek Inc.: MediaTek offers automotive-grade MCUs optimized for various applications, including engine control, infotainment, and driver assistance systems. Its MCUs provide high performance and reliability for Taiwanese and global automakers.

Graphics Processing Units (GPUs):

Key Players:

Huawei Technologies (HiSilicon): Huawei's semiconductor subsidiary, HiSilicon, develops advanced GPUs for automotive infotainment and navigation systems. Its solutions provide high-quality graphics rendering and multimedia capabilities for Chinese vehicles.

Rockchip Electronics Co., Ltd.: Rockchip is a leading Chinese semiconductor company specializing in multimedia and graphics processing solutions. Its GPUs are used in automotive infotainment systems, offering immersive user experiences.

MediaTek Inc.: MediaTek develops GPU solutions for automotive infotainment systems, navigation, and multimedia applications. Its GPUs deliver high-quality graphics rendering and multimedia performance for Taiwanese vehicles

Digital Signal Processors (DSPs):

Key Players:

MediaTek Inc.: MediaTek develops DSP solutions for automotive audio processing, voice recognition, and sensor data processing. Its DSPs are integrated into Chinese vehicles for enhanced audio quality and connectivity features.

Spreadtrum Communications (Unisoc): Unisoc, formerly Spreadtrum Communications, provides DSP solutions optimized for automotive applications. Its DSPs enable real-time signal processing and sensor data fusion in Chinese vehicles.

Realtek Semiconductor Corp.: Realtek provides DSP solutions for automotive audio processing, voice recognition, and connectivity. Its DSPs enhance audio quality and support hands-free communication in Taiwanese vehicles.

Field-Programmable Gate Arrays (FPGAs):

Key Players:

Alibaba Group (T-Head Semiconductor): T-Head Semiconductor, a subsidiary of Alibaba Group, develops FPGA solutions for automotive prototyping and customization. Its FPGAs enable rapid system development and customization for Chinese automotive OEMs.

ZTE Corporation: ZTE offers FPGA solutions tailored for automotive applications, including sensor interfacing, data preprocessing, and in-vehicle networking. Its FPGAs provide flexibility and programmability for Chinese automakers.

Alchip Technologies Ltd.: Alchip specializes in FPGA design and customization for automotive applications. Its FPGAs enable rapid prototyping and customization for Taiwanese automotive OEMs.

Application-Specific Integrated Circuits (ASICs):

Key Players:

Huawei Technologies (HiSilicon): HiSilicon designs ASICs optimized for specific automotive applications, such as ADAS and in-vehicle networking. Its custom ASIC solutions are integrated into Chinese vehicles for enhanced safety and connectivity.

Baidu Inc. (Kunlun): Baidu's semiconductor subsidiary, Kunlun, develops AI-focused ASICs for autonomous driving and AI-powered applications in vehicles. Its ASIC solutions are deployed in Chinese autonomous vehicles and smart transportation systems.

Global Unichip Corp. (GUC): GUC offers custom ASIC solutions for automotive applications, including ADAS, in-vehicle networking, and infotainment systems. Its ASICs provide high performance and reliability for Taiwanese vehicles.

Memory Chips:

Key Players:

Yangtze Memory Technologies Co., Ltd. (YMTC): YMTC is a leading Chinese memory chip manufacturer, producing Flash memory and

DRAM solutions for automotive applications. Its memory chips support OTA updates, vehicle diagnostics, and multimedia storage in Chinese vehicles.

Innotron Memory (Hefei Chang Xin): Innotron Memory specializes in DRAM production for various industries, including automotive. Its memory chips provide high-speed data storage and reliability for Chinese automotive electronics.

Winbond Electronics Corp.: Winbond produces Flash memory and DRAM solutions for automotive electronics. Its memory chips support data storage, firmware updates, and multimedia applications in Taiwanese vehicles.

Sensors and Sensor Interface Chips:

Key Players:

Huawei Technologies: Huawei develops automotive sensors and sensor interface chips for Chinese vehicles, supporting applications such as navigation, motion detection, and driver assistance systems.

Bosch (Bosch Automotive Products (Suzhou) Co., Ltd.): Bosch, a global leader in automotive sensors, has a significant presence in China. Its sensor solutions are widely used in Chinese vehicles for safety, efficiency, and comfort applications.

Delta Electronics Inc.: Delta Electronics develops automotive sensors and interface chips for various applications, including environmental monitoring, driver assistance, and safety systems. Its solutions enhance vehicle safety and efficiency for Taiwanese automakers.

Power Management ICs:

Key Players:

Semiconductor Manufacturing International Corporation (SMIC): SMIC is China's largest semiconductor foundry, providing power management ICs for automotive applications. Its solutions ensure stable power supply and energy efficiency in Chinese vehicles.

Huawei Technologies: Huawei develops power management ICs for automotive systems, including battery management, motor control, and energy efficiency optimization. Its solutions contribute to the reliability and performance of Chinese vehicles.

Maxim Integrated Products, Inc. (Taiwan Branch): Maxim Integrated offers power management ICs for automotive systems, including battery management, motor control, and energy efficiency optimization. Its solutions ensure reliable power supply and efficiency for Taiwanese vehicles.

South Korea:

Microcontrollers (MCUs):

Samsung Electronics Co., Ltd.: Samsung provides automotive-grade MCUs for various applications, including powertrain control, infotainment, and vehicle networking. Its MCUs offer high performance and reliability for South Korean and global automakers.

Graphics Processing Units (GPUs):

Samsung Electronics Co., Ltd.: Samsung develops GPU solutions for automotive infotainment systems, navigation, and driver assistance features. Its GPUs deliver immersive graphics rendering and multimedia experiences for South Korean vehicles.

Digital Signal Processors (DSPs):

LG Electronics Inc.: LG Electronics offers DSP solutions for automotive audio processing, voice recognition, and connectivity. Its DSPs enhance audio quality and support advanced communication features in South Korean vehicles.

Field-Programmable Gate Arrays (FPGAs):

SK hynix Inc.: SK hynix specializes in FPGA solutions for automotive applications, including sensor interfacing, data preprocessing, and in-vehicle networking. Its FPGAs provide flexibility and customization for South Korean automakers.

Application-Specific Integrated Circuits (ASICs):

Samsung Electronics Co., Ltd.: Samsung designs custom ASICs for automotive applications, including safety-critical systems and advanced driver assistance features. Its ASIC solutions offer high performance and reliability for South Korean vehicles.

Memory Chips:

SK hynix Inc.: SK hynix produces Flash memory and DRAM solutions for automotive electronics. Its memory chips support data storage, firmware updates, and multimedia applications in South Korean vehicles.

Sensors and Sensor Interface Chips:

LG Innotek Co., Ltd.: LG Innotek develops automotive sensors and interface chips for various applications, including camera systems, radar sensors, and environmental monitoring. Its solutions enhance safety and efficiency in South Korean vehicles.

Power Management ICs:

Magnachip Semiconductor Corp.: Magnachip offers power management ICs for automotive systems, including voltage regulation, power distribution, and energy efficiency optimization. Its solutions ensure reliable power supply and efficiency for South Korean vehicles.

Including these companies from Taiwan and South Korea showcases the diverse landscape of the automotive semiconductor market, highlighting the contributions of these key players to the advancement of automotive technology and innovation.

Conclusion

As the automotive industry continues its evolution towards software-defined vehicles, the chip-to-cloud strategy will play a pivotal role in shaping the future of transportation. By leveraging the combined power of onboard chips and cloud computing, automakers can unlock new levels of innovation, efficiency, and safety. However, to fully realize the potential of this approach, collaboration between industry stakeholders, regulatory bodies, and technology providers will be essential. With the right strategy and mindset, the journey towards a connected, intelligent automotive ecosystem promises to revolutionize the way we experience mobility.

NINE

ROLE OF CLOUD PROVIDERS

As the automotive industry embraces the era of software-defined vehicles (SDVs), the integration of advanced technology into vehicles necessitates a paradigm shift in vehicle architecture and management. One of the key elements in this transition is the utilization of cloud services, which play a crucial role in managing complex vehicle software architectures securely and efficiently. In this article, we will explore how cloud providers are poised to shape the future of SDVs and highlight key players in both the global and Chinese markets.

The Evolution of Software-Defined Vehicles:

Software-defined vehicles represent a monumental shift in the automotive landscape, where software components take precedence over traditional hardware-centric systems. This shift enables vehicles to become more adaptable, intelligent, and connected, offering enhanced functionality, customization, and user experiences.

However, the increasing complexity of software architectures within vehicles poses significant challenges in terms of management, security, and scalability. To address these challenges, automotive manufacturers are turning to cloud computing solutions to centralize and streamline software management processes.

The Role of Cloud Providers:

Cloud providers offer a range of services and solutions tailored to the unique requirements of the automotive industry. These include:

Data Storage and Processing: Cloud platforms provide scalable storage solutions and powerful processing capabilities, allowing automotive manufacturers to store and analyze vast amounts of vehicle data generated from sensors, cameras, and other sources.

Software Updates and Management: Cloud-based software deployment and management platforms enable seamless over-the-air (OTA) updates, allowing manufacturers to deliver software patches, bug fixes, and new features to vehicles remotely.

Security and Compliance: Cloud providers offer robust security measures and compliance frameworks to safeguard vehicle data and ensure regulatory compliance. This includes encryption, authentication, and access control mechanisms to protect sensitive information.

Fleet Management and Monitoring: Cloud-based fleet management platforms enable real-time monitoring of vehicle performance, health status, and diagnostic data, allowing manufacturers to proactively identify and address issues to ensure optimal vehicle operation.

Key Cloud Players in the Global Market:

Amazon Web Services (AWS): AWS offers a comprehensive suite of cloud services, including Amazon EC2, Amazon S3, and AWS IoT, tailored to the automotive industry. Its scalable infrastructure and AI/ML capabilities enable automotive manufacturers to innovate rapidly and securely.

Microsoft Azure: Microsoft Azure provides a range of cloud services, including Azure IoT Hub, Azure Machine Learning, and Azure Security Center, designed to support the development and deployment of SDVs. Its robust security features and global presence make it a preferred choice for automotive OEMs.

Google Cloud Platform (GCP): GCP offers advanced analytics, AI, and machine learning solutions, such as Google Cloud AutoML and TensorFlow, to enable automotive manufacturers to derive insights from vehicle data and enhance driver experiences. Its data storage and processing capabilities are well-suited for SDV applications.

Key Cloud Players in the Chinese Market:

Alibaba Cloud: Alibaba Cloud is a leading cloud service provider in China, offering a wide range of services, including Alibaba Cloud IoT, Alibaba Cloud ECS, and Alibaba Cloud Security, tailored to the automotive industry. Its strong presence in the Chinese market makes it a preferred choice for domestic automotive manufacturers.

Tencent Cloud: Tencent Cloud provides cloud services, such as Tencent Cloud IoT and Tencent Cloud Virtual Machines, designed to support SDV development and deployment in China. Its expertise in AI and data analytics enables automotive OEMs to harness the power of vehicle data for insights and innovation.

Huawei Cloud: Huawei Cloud offers a comprehensive portfolio of cloud services, including Huawei Cloud EI, Huawei Cloud ECS, and Huawei Cloud Security, specifically designed for the automotive sector. Its focus on security, reliability, and performance makes it a trusted partner for automotive companies in China.

Baidu Cloud: Baidu Cloud is a prominent cloud service provider in China, offering a range of services, including Baidu Cloud AI, Baidu Cloud ECS, and Baidu Cloud Security, tailored to the needs of the automotive industry. Its expertise in artificial intelligence and data analytics enables automotive manufacturers to leverage advanced technologies for SDV development and deployment.

How do you bring Cloud Native Development practices to the embedded edge

Bringing cloud-native development practices to the embedded edge involves adapting established methodologies and technologies from cloud computing to the constraints and requirements of edge computing environments. Here's a step-by-step guide on how to achieve this:

Understand Embedded Edge Constraints: Begin by understanding the unique constraints of embedded edge devices, such as limited processing power, memory, and energy resources. Unlike cloud environments, embedded edge devices operate with restricted hardware capabilities, making it essential to optimize resource usage.

Select Suitable Hardware: Choose hardware platforms that are compatible with cloud-native development practices and capable of supporting containerization and orchestration technologies. Look for embedded systems with sufficient processing power, memory, and storage capacity to run containerized applications efficiently.

Containerize Applications: Containerization is a fundamental aspect of cloud-native development. Adapt your embedded edge applications to run within lightweight, isolated containers using containerization platforms like Docker or containerd. This enables greater portability, scalability, and flexibility in deploying and managing applications across diverse edge environments.

Implement Orchestration: Employ orchestration frameworks such as Kubernetes or Docker Swarm to manage containerized applications at the edge. These platforms automate deployment, scaling, and monitoring tasks, ensuring efficient resource utilization and high availability of edge services. Kubernetes, in particular, offers features like edge node support and lightweight distributions tailored for edge computing.

Optimize Networking: Address network connectivity challenges inherent in edge environments by optimizing networking configurations. Implement edge-native networking solutions that prioritize low latency, reliability, and bandwidth efficiency. Technologies like edge computing gateways and mesh networking can facilitate seamless communication between edge devices and cloud services.

Enable Edge Intelligence: Leverage cloud-native development practices to enable edge intelligence and real-time data processing at the edge. Utilize edge computing frameworks and machine learning models to perform data analytics, inference, and decision-making directly on embedded devices. This minimizes data transfer to the cloud, reduces latency, and enhances responsiveness in edge applications.

Implement DevOps Processes: Adopt DevOps methodologies and practices to streamline the development, deployment, and maintenance of embedded edge applications. Implement continuous integration and continuous deployment (CI/CD) pipelines tailored for edge environments to automate testing, packaging, and deployment workflows. Embrace infrastructure as code (IaC) principles to manage edge infrastructure programmatically.

Ensure Security and Compliance: Prioritize security and compliance considerations when implementing cloud-native development practices at the embedded edge. Implement security best practices, such as secure boot, encryption, and access control mechanisms, to protect edge devices and data from threats and vulnerabilities.

Ensure compliance with industry regulations and standards relevant to edge computing and data privacy.

Monitor and Manage Edge Infrastructure: Implement monitoring and management solutions to oversee the health, performance, and security of edge infrastructure and applications. Utilize edge-native monitoring tools and centralized management platforms to gain visibility into edge deployments, troubleshoot issues, and enforce governance policies.

Iterate and Improve: Continuously iterate and improve your cloud-native development practices for the embedded edge based on feedback, performance metrics, and evolving requirements. Embrace a culture of experimentation and innovation to drive ongoing optimization and adaptation to changing edge computing landscapes.

Conclusion

As the automotive industry transitions to software-defined vehicles, cloud providers play a pivotal role in enabling the development, deployment, and management of next-generation vehicle platforms and components. With their scalable infrastructure, advanced analytics capabilities, and robust security measures, cloud platforms empower automotive manufacturers to innovate rapidly, enhance customer experiences, and drive the future of mobility.

By leveraging cloud services from key players in both the global and Chinese markets, automotive OEMs can navigate the complexities of SDV development and emerge as leaders in the era of connected, intelligent transportation.

TEN

HOW ELECTROMOBILITY IS TRANSFORMING BY SOFTWARE DEFINED VEHICLES

The electric vehicle (EV) revolution is gaining momentum, but the future of electromobility lies not just in powerful batteries and sleek designs, but in the software that underpins them. Software-defined vehicles (SDVs) are fundamentally changing the landscape, transforming EVs from mere electric cars into rolling ecosystems of connected experiences.

Unlocking Battery Potential: From Pack to Powerhouse

Battery management systems (BMS) are the unsung heroes of EVs. Traditionally, BMS software focused primarily on safety and basic cell health monitoring. However, software advancements are unlocking new possibilities:

Optimizing Range and Performance: SDVs can analyze driving patterns and real-time conditions to adjust charging and power delivery for maximum efficiency and extended range. Companies like Tesla and General Motors are leading the charge in developing in-house battery management software.

Personalized Charging Strategies: Software can learn from driver behavior and recommend optimal charging schedules based on daily commutes and long trips. This not only optimizes battery life but also minimizes charging time and cost.

Chips Take the Lead: Programmable Power

The brains of an SDV are the powerful chips controlling everything from in-vehicle entertainment to driver assistance systems. Here's where software shines:

Over-the-Air (OTA) Updates: Software updates can be delivered wirelessly, allowing automakers like Volkswagen and Bosch to continuously improve vehicle performance, address bugs, and even unlock new features after purchase. This keeps EVs feeling fresh and adaptable to evolving technology.

Customizable Driving Experiences: Imagine adjusting steering sensitivity or throttle response through software settings. SDVs enable personalization, allowing drivers to tailor their experience based on preference and road conditions. Companies like Continental and Aptiv are at the forefront of developing software-defined driving experiences.

Charging Infrastructure: Smarter Than Ever

The charging infrastructure plays a crucial role in EV adoption. Here's how software is making a difference:

Real-time Monitoring and Diagnostics: Software platforms can track charging station availability, power output, and potential maintenance needs. This information can be accessed by drivers and station operators, ensuring a smooth and efficient charging experience. Companies like ABB and Siemens are leaders in developing smart charging infrastructure solutions.

Demand Management and Grid Integration: Software can analyze charging patterns and optimize energy usage to avoid overloading the grid. This is crucial with large-scale EV adoption and the need for sustainable energy solutions. Companies like Enel X and ChargePoint are making strides in smart grid integration for electromobility.

The Road Ahead: Collaboration is Key

The transformation of electromobility requires collaboration between traditional car manufacturers (OEMs) and Tier 1 suppliers. Here are some key players leading the charge:

OEMs: Tesla, General Motors, Volkswagen, and Ford are actively developing in-house software capabilities and forging partnerships with technology companies.

Tier 1 Suppliers: ZF ,Bosch, Continental, Aptiv, and Magna are leveraging their expertise in hardware and software to provide holistic solutions for SDVs and charging infrastructure.

The Future is Electric, and it's Powered by Software

Software-defined vehicles are revolutionizing electromobility, pushing the boundaries of battery performance, chip capabilities, and charging infrastructure. With continued advancements and collaboration, EVs are poised to become not just cleaner alternatives, but intelligent and adaptable companions on the road to a sustainable future.

ELEVEN

HOW SOFTWARE-DEFINED VEHICLES (SDVS) DRIVE COST OPTIMIZATION FOR CARMAKERS

How Software-Defined Vehicles (SDVs) Drive Cost Optimization for Carmakers

The automotive industry is undergoing a seismic shift with the rise of software-defined vehicles (SDVs). These vehicles prioritize software as the core component, allowing for continuous improvement, feature upgrades, and customization – all through over-the-air updates. This new approach presents a compelling opportunity for carmakers to achieve significant cost optimization in the long term.

Traditional Challenges and the Rise of Complexity

Traditionally, car manufacturing has been a complex and expensive process. Vehicles are intricate puzzles of hardware components, each requiring extensive research, development, and sourcing. Modifications and feature updates necessitate physical changes to the vehicle, driving up production costs and extending time to market for new features.

This complexity is further amplified with the rapid pace of innovation, especially in areas like driver assistance systems and connectivity features. Carmakers struggle to keep pace with the latest technology, often resulting in vehicles with outdated features within a few years of production.

SDVs: Streamlining Complexity, Reducing Costs

SDVs offer a compelling solution to these challenges. By separating hardware and software, carmakers can focus on developing a robust and flexible hardware platform. This platform can then accommodate a wide range of software functionalities, allowing for customization and feature upgrades without physical modifications.

This approach offers several cost-saving advantages:

Reduced Hardware Variation: With a standardized hardware platform, carmakers can streamline production lines and reduce the need for multiple vehicle variants based solely on software features. This simplifies production processes and potentially lowers manufacturing costs.

Shorter Development Cycles: Software updates can be deployed quickly over-the-air, eliminating the need for lengthy retooling or production line changes. This allows carmakers to introduce new features and functionalities faster, keeping pace with market demands.

Improved Inventory Management: Standardized hardware platforms make it easier to manage parts inventory. This reduces storage costs and the risk of obsolete parts accumulating.

Beyond Production: Long-Term Cost Benefits

The benefits of SDVs extend beyond the production line. Here are some additional ways carmakers can achieve cost optimization in the long term:

Software-Based Diagnostics and Maintenance: SDVs can provide real-time data on vehicle performance and potential issues. This allows for preventative maintenance and remote diagnostics, reducing the need for costly repairs and improving vehicle uptime for fleets or ride-sharing services.

Subscription-Based Revenue Models: Software-defined features can be offered as subscriptions, creating a recurring revenue stream for carmakers. This allows them to monetize new features and functionalities without relying on hardware upgrades.

Improved Efficiency and Performance: Software updates can optimize engine performance, improve fuel efficiency, and enhance safety features. This translates to lower operating costs for both carmakers and vehicle owners.

Challenges and Considerations

While the potential for cost optimization is significant, transitioning to SDVs is not without challenges. Carmakers need to invest heavily in software development capabilities, cybersecurity measures, and over-the-air update infrastructure. Additionally, ensuring a smooth user experience with software updates and potential compatibility issues requires careful planning and execution.

The Road Ahead: A Software-Defined Future

Despite the challenges, the long-term cost benefits and market potential make the shift towards SDVs an attractive proposition for carmakers. By embracing software as the core, car manufacturers can streamline production, reduce costs, and deliver a more adaptable and user-centric driving experience. The future of the automotive industry is likely software-defined, and carmakers that successfully navigate this transition will be well-positioned for long-term success.

PART 2

The Business of Software Defined Vehicles

TWELVE

CHINA'S AUTOMOTIVE JOURNEY

I must begin writing about the business of SDVs from China's Automotive Journey.

China has experienced a dramatic transformation in its automotive industry. I vividly recall my first visit in 2009 as part of the IIM Bangalore Exchange program with Tshinghua University. At that time, China's car sales were hovering around 8 million units, and many experts predicted that, over the next decade, the market might grow to 16 million or possibly 20 million units. However, to everyone's surprise, China surpassed the 20 million mark by 2015, shattering all predictions, including those from McKinsey and other major consulting firms. This was a clear indication of the exponential growth China was experiencing in an industry that often serves as a leading indicator of the economy.

If I were to break down China's automotive development into phases, I would describe it as evolving in three distinct stages.

Phase 1 (2009–2015): The first phase was characterized by joint ventures between state-owned Chinese companies and global automotive players. During this time, Chinese manufacturers began developing their own technologies, producing replicas of the cars made under these joint ventures. For instance, when SAIC Motor, a Chinese state-owned enterprise, formed a joint venture with Volkswagen, Volkswagen cars featured electronic power steering (EPS), but SAIC's own cars lacked this feature. SAIC quickly embarked on a journey to develop their own EPS in-house, initially using hydraulic power steering but ultimately moving towards electronic systems. This sparked a significant shift, with Chinese manufacturers not only creating their own components but also developing a network of Tier 1 suppliers and OEMs to support their ecosystems. This laid the foundation for the backbone of China's automotive industry.

Phase 2 (2015–2021): The second phase saw a surge in the number of startups entering the electric vehicle (EV) space, alongside small provincial car manufacturers. Prominent names such as NIO, which set up its plant in Shanghai and an R&D center in Hefei, emerged during this period. Companies like BYD, Changan, and Great Wall Motors also gained traction, introducing multiple car models with enhanced features. This phase marked the democratization of the automotive ecosystem, as local manufacturers expanded and began offering a wider variety of cars at competitive prices, each with an increasing range of advanced features.

Phase 3 (2022 and beyond): From 2022 onwards, we entered the third phase, where Chinese state-owned companies, once dependent on foreign joint venture partnerships, began producing cars that were far superior to their global counterparts. This marked a game-changing shift. Local manufacturers began to outpace foreign multinationals in both quality and innovation. Over a span of just 3-4 years, the sales volume of Chinese car makers surged, while that of multinational corporations declined. This trend suggests that, in the next decade, it will become exceedingly difficult for foreign automakers to compete in China. The combination of lower production costs, advanced features, and cutting-edge innovations offered by local manufacturers will create an insurmountable competitive advantage. Much like Japan, where it is tough for foreign automakers to enter the market, China's domestic players—brands like BYD and Geely—are quickly becoming the Toyota and Honda of the future.

In conclusion, China's automotive landscape is undergoing a major transformation. The market dominance of local manufacturers will continue to rise, and by 2035, it may be nearly impossible for foreign players to enter the Chinese market effectively. The pace of innovation and cost competitiveness from domestic manufacturers is set to reshape the global automotive industry.

THIRTEEN

How Software is Driving Automotive Company Valuations

In the past, the value of an automotive company was measured by the strength of its manufacturing assets, global footprint, and vehicle volumes. Brands were built on horsepower, design, and engine reliability. But that era is fading fast. In today's market, companies that embrace software-defined vehicle (SDV) architectures are not just delivering better customer experiences— they are achieving better sales growth and commanding higher valuations.

Tesla was the first to flip the narrative. Traditional OEMs used to scoff at its lack of production heritage, yet today, Tesla's valuation has surpassed the combined value of multiple legacy automakers. Why? Because Tesla redefined the car as a software platform on wheels—capable of continuous updates, data-driven personalization, and even post-sale monetization.

This shift isn't just about stock market hype. It reflects a deeper structural change in how value is created in the automotive industry. Legacy OEMs are weighed down by rigid hardware-software integration, siloed ECUs, and complex supplier dependencies. In contrast, SDV-focused players are streamlining architectures, centralizing computing, and leveraging cloud-native practices like containerization, OTA updates, and microservices to innovate faster and reduce lifecycle costs.

Recent data supports this trend. Companies investing aggressively in software platforms—like Volkswagen Group with CARIAD, or GM with Ultifi—are seeing higher software-related revenue, stronger customer retention, and better adaptability to new mobility trends (e.g., EVs, shared services). Analysts now look not just at units sold, but at the "lifetime software revenue per vehicle"—a metric that didn't even exist a decade ago.

Moreover, SDV-aligned companies are attracting better tech partnerships, faster developer ecosystems, and premium customer segments willing to pay for digital features. This unlocks new recurring revenue streams—from navigation and infotainment to advanced driver-assistance systems (ADAS), autonomous features, and vehicle personalization.

In a world where hardware is getting commoditized and electrification levels the playing field, software is the real differentiator. Automotive companies that don't make this pivot risk falling behind—not just in innovation, but in how investors perceive their future growth and profitability.

Valuation is no longer tied only to the number of cars you build—it's tied to how intelligent, upgradeable, and monetizable those cars become after they leave the factory.

It's not just a technological evolution. It's a business transformation—from hardware to code, from one-time sale to lifetime value.

FOURTEEN

THE FUTURE OF OEM-TIER 1 COLLABORATION

The rise of software-defined vehicles (SDVs) is fundamentally changing the way Original Equipment Manufacturers (OEMs) and Tier 1 suppliers collaborate. This chapter explores the evolving landscape of partnerships, the rise of open-source platforms, and the impact on the traditional automotive supply chain.

New Business Models and Partnerships in the SDV Era

Traditional car manufacturing relied heavily on Tier 1 suppliers providing complete hardware modules. In the SDV era, the lines are blurring. Here are some emerging business models:

Joint software development: OEMs and Tier 1s are collaborating to develop core software components like operating systems, middleware, and specific applications. This requires a shift from hardware-centric partnerships to joint software ownership and development.

Tier 1 specialization: Tier 1s are increasingly specializing in specific software domains like ADAS, connectivity, or in-vehicle experiences. This allows them to offer expertise and pre-built software modules to OEMs.

Service-oriented partnerships: Tier 1s may transition from selling hardware to offering software-as-a-service (SaaS) solutions. This allows for continuous innovation and revenue streams based on ongoing software updates and feature enhancements.

Platform partnerships: Open-source platforms or consortium-developed platforms could emerge, where both OEMs and Tier 1s contribute and leverage core software components. This fosters collaboration and reduces redundancy in development efforts.

Open-Source Platforms and Collaborative Development

The traditional closed nature of automotive software development is giving way to a more open approach. Here's how open-source platforms can play a role:

Faster innovation: Open-source platforms allow for faster development cycles by leveraging a wider pool of developers and code contributions.

Reduced development costs: Sharing codebases and functionalities reduces duplication of effort and lowers development costs for all participants.

Standardization and interoperability: Open-source platforms can promote common standards and APIs, facilitating easier integration of components from different vendors.

Security benefits: A larger community can identify and address security vulnerabilities more efficiently in open-source software.

However, open-source adoption in the automotive industry also faces challenges:

Safety concerns: Critical software functions in vehicles require rigorous testing and certification, which might be more challenging in open-source environments.

Intellectual property (IP) protection: OEMs may be hesitant to share sensitive code related to differentiation features.

The Changing Dynamics of the Automotive Supply Chain

The rise of SDVs is reshaping the traditional automotive supply chain:

Vertical integration: OEMs might choose to develop more software in-house, potentially reducing reliance on Tier 1s for specific functionalities.

Tier 1 consolidation: As software becomes a core competency, Tier 1s with strong software expertise might consolidate their positions.

New players: Tech companies and startups with software expertise are entering the automotive space, potentially disrupting traditional supplier relationships.

Tiered supplier structure: A multi-tiered supplier landscape might emerge, with specialized software vendors working alongside traditional hardware suppliers.

The future of OEM-Tier 1 collaboration in the SDV era will likely involve a mix of these models. The most successful partnerships will be those that embrace flexibility, foster innovation, and adapt to the rapidly changing technological landscape.

FIFTEEN

VENTURE CAPITALISTS AND STARTUPS IN THE SOFTWARE-DEFINED VEHICLE (SDV) SPACE

The rise of SDVs is attracting significant investment, with both established VCs and innovative startups shaping the future of mobility. Let's leverage Crunchbase data to explore some key players:

USA

Venture Capitalists:

Early Stage Investors:

Trucks Venture Capital (San Francisco): Focuses on early-stage transportation investments. According to Crunchbase, they have invested in companies like Aurora (self-driving software) and May Mobility (autonomous vehicle technology).

Maven Ventures (Los Angeles): Invests in high-growth software startups. While specific names aren't publicly available, Crunchbase suggests they may have investments in early-stage SDV companies based on their focus on transportation and software.

Right Side Capital Management (San Francisco): Invests in pre-seed technology startups across various sectors, including automotive. Crunchbase search results don't definitively show SDV investments, but you can explore their portfolio companies for potential leads.

Later Stage Investors:

Sequoia Capital (Global): A renowned VC firm with a history of backing tech giants. Crunchbase data confirms their investment in Aurora, highlighting their interest in the open-source approach to SDV software.

Andreessen Horowitz (Silicon Valley): Another prominent VC heavily invested in the future of mobility. According to Crunchbase, they've backed Aurora and Luminar, a leading lidar company (critical sensor for autonomous vehicles) – signifying their multi-faceted approach to the SDV space.

SoftBank Vision Fund (Global): A massive investment fund. While not all investments are publicly disclosed, Crunchbase data reveals their backing of Cruise (self-driving car company acquired by General Motors), demonstrating their commitment to the SDV revolution.

Startups:
Autonomous Driving:

Aurora (California): Developing self-driving software using an open-source approach. Crunchbase data provides a detailed profile of Aurora, including funding rounds, investors, and team information.

Waymo (California): A self-driving car company owned by Google's parent company, Alphabet. While not a typical startup due to its ownership structure, Waymo's pioneering role in the space can't be ignored.

Cruise (California): Acquired by General Motors, Cruise focuses on self-driving vehicle development and testing. Crunchbase offers insights into their funding history and key personnel.

In-Vehicle Software and Connectivity:

Apple (California): CarPlay is their in-vehicle infotainment system designed for seamless integration with Apple devices. Crunchbase isn't applicable here as Apple is a publicly traded company. However, industry news and reports can be found to understand their SDV strategy.

Android (Global, owned by Google): Android Automotive OS is an open-source platform for in-vehicle infotainment. Similar to Apple, you can't use Crunchbase for Google, but industry resources provide insights into their SDV approach.

HERE Technologies (Netherlands): Provides mapping and location data services for vehicles. Crunchbase offers a profile of HERE Technologies, including their funding history, products, and market position within the SDV ecosystem.

Venture Capitalists:
China:

Wu Capital (China): Focuses on artificial intelligence (AI) and future mobility. According to Crunchbase, they've invested in NIO Capital, the venture arm of a leading Chinese electric vehicle (EV) maker. While not a direct SDV investment, it indicates their interest in the electric and potentially software-defined future of transportation.

Jeneration Capital (China): Invests in deep technology and autonomous driving. Crunchbase data shows their backing of companies like WeRide, a Chinese self-driving car startup.

NIO Capital (China): The venture capital arm of NIO, a prominent Chinese EV manufacturer. While not a typical VC firm, their focus on EVs positions them strategically in the SDV space.

India:

Indus Capital Partners (India): Invests in growth-stage companies across sectors. Crunchbase data reveals their investment in Ola Electric, a manufacturer of electric scooters. While not purely SDV, it highlights their interest in electric mobility, which often overlaps with SDVs.

Blume Ventures (India): Focuses on technology investments. According to Crunchbase, they've backed FleetX, a company providing electric vehicle logistics solutions. This investment showcases their interest in software-enabled solutions within the electric vehicle space.

Kalaari Capital (India): Invests in early-stage technology startups. While Crunchbase doesn't show specific SDV investments, their portfolio includes companies working on connected vehicle technologies, which could position them well for the SDV future.

Startups:
China:

Baidu Apollo (Beijing): Develops autonomous driving technology. Crunchbase provides a profile of Baidu Apollo, detailing their funding and partnerships.

WeRide (Guangzhou): A self-driving car startup. According to Crunchbase, they've received funding from prominent VCs like Jeneration Capital.

Pony.ai (California & China): Provides autonomous driving tech for robotaxis. Crunchbase data showcases their global presence and investments received.

India:

SatNav Technologies (Bengaluru): Develops GPS navigation and fleet management solutions. While not directly focused on autonomous driving, Crunchbase information highlights their role in the connected vehicle space, which can be a stepping stone towards SDVs.

Tork Motors (Pune): Manufactures electric motorcycles. Crunchbase data provides insights into their funding and product development. Electric vehicles are prime candidates for software-defined features, making Tork Motors a potential player in the SDV space.

Ather Energy (Bengaluru): Manufactures electric scooters with connected vehicle tech. Crunchbase offers information about their funding and market position. Their focus on electric vehicles with connected features positions them well for the SDV future.

SIXTEEN

The Impact of SDVs on Insurance Companies: A Paradigm Shift on the Horizon

The Impact of SDVs on Insurance Companies: A Paradigm Shift on the Horizon

The rise of software-defined vehicles (SDVs) and the potential for autonomous driving are poised to disrupt the insurance industry in profound ways. These advancements will not only change the nature of car ownership but also fundamentally alter the risk profile associated with operating a vehicle. As a result, leading insurance companies around the globe need to be prepared to adapt their business models to this evolving landscape.

Shifting Ownership Models: From Owning to Accessing

The convenience and affordability of ride-sharing services like Uber and Didi Chuxing (China) and car-subscription models offered by companies like General Motors' Maven and Zoomcar (India) are already impacting traditional car ownership. With SDVs, this trend is likely to accelerate. Consumers may prioritize access to mobility over vehicle ownership, leading to a decline in private auto insurance policies. Insurance companies will need to develop innovative products tailored to these new ownership models, like usage-based insurance that charges based on miles driven or time spent in the vehicle, already being offered by Progressive with their "Pay As You Drive" program.

The Rise of Autonomous Vehicles: Reshaping Risk

Autonomous vehicles, pioneered by companies like Tesla (US), Waymo (owned by Alphabet, US), and Cruise (owned by General Motors, US) in the West, are being paralleled by efforts from Eastern companies like Baidu (China) and Mahindra & Mahindra (India). These advancements promise a future with significantly fewer accidents, a major concern for insurance companies. With software-controlled driving and advanced safety features, the risk of human error – a major contributor to accidents – will be drastically reduced. This could lead to a significant decrease in insurance premiums, potentially impacting current pricing models based on factors like driver history and age.

New Liabilities: Shifting Responsibility

The transition to autonomous vehicles raises questions about liability in case of an accident. Will the responsibility fall on the manufacturer, like Ford (US) or Toyota (Japan), the software developer, like NVIDIA (US) or Mobileye (Israel), or the passenger? Insurance companies will need to develop new types of coverage to address these evolving liabilities. This could involve insuring the vehicle itself, the software systems, or even offering product liability coverage to manufacturers.

The Data Advantage: Personalized Risk Assessment

SDVs will generate a wealth of data on driving behavior, vehicle performance, and environmental conditions. This data can be used by insurance companies like Nationwide (US), The Hartford (US), Ping An Insurance (China), and Tokio Marine (Japan) to create personalized risk profiles for each individual car or driver. This granular data analysis allows for more accurate risk assessment, potentially leading to fairer premiums based on individual driving habits and safety records.

A Call for Innovation and Collaboration

The impact of SDVs on the insurance industry requires adaptation and innovation. Collaboration between automakers, technology companies, and insurance providers will be crucial in developing new risk assessment tools, insurance products, and regulatory frameworks. For instance, a partnership between State Farm (US) and Waymo (US) could lead to the development of specialized insurance policies for autonomous vehicles.

Looking Forward: Embracing the Change

The future of mobility is likely to be dominated by SDVs and autonomous vehicles. While the disruption to the insurance industry may be significant, it also presents opportunities. By embracing these changes, developing new insurance products, and utilizing data effectively, insurance companies can remain relevant and play a critical role in this evolving mobility landscape.

The road ahead will be paved with innovation and adaptation. It is the companies that can anticipate and navigate this shift that will thrive in the era of software-defined vehicles and autonomous driving.

SEVENTEEN

THE BATTLE FOR CONTROL: AUTOMAKERS VS. TECH COMPANIES IN THE SDV ERA

The software-defined vehicle (SDV) revolution is transforming the automotive industry, blurring the lines between traditional car manufacturers and tech giants. This shift raises a critical question: who will control the future of SDVs? Will established automakers leverage their engineering prowess and brand legacy, or will tech companies capitalize on their software expertise and user experience mastery?

Traditional Automakers: A Legacy of Hardware

Automakers boast a long history of building reliable and safe vehicles. They possess deep engineering expertise in areas like powertrain development, chassis design, and safety systems. Additionally, established brands hold a strong reputation and customer loyalty, giving them a significant advantage in the market.

However, automakers face challenges adapting to the software-centric nature of SDVs. Their traditional development cycles are slow, and their corporate cultures may not be as agile as those of tech companies. Furthermore, they may lack the in-house software development expertise needed to compete effectively.

Tech Companies: A Software Advantage

Tech companies bring a wealth of software development experience to the table. They excel in user interface (UI) design, cloud computing, and artificial intelligence (AI). This expertise allows them to create intuitive in-vehicle experiences and develop advanced features like autonomous driving functionalities. Additionally, tech companies are known for their rapid innovation cycles and their ability to disrupt established industries.

Despite their strengths, tech companies lack the experience of traditional automakers in vehicle engineering and safety. Partnering with established manufacturers to gain this expertise is crucial for their success in the SDV market. Additionally, building trust with consumers who may be wary of a tech company controlling their car could be a significant hurdle.

The Potential for Collaboration

The future of SDVs might not be a winner-take-all scenario. Collaboration between automakers and tech companies seems more likely. Automakers can leverage their hardware expertise and brand reputation, while tech companies can contribute their software development prowess and user-centric approach.

Strategic partnerships would allow both sides to address their weaknesses. Automakers could gain access to cutting-edge software and tech talent, while tech companies could benefit from the manufacturing expertise and established distribution channels of traditional car manufacturers.

Who Will Prevail? It's Still Unclear

The battle for control of the SDV market is far from over. Both automakers and tech companies have their strengths and weaknesses.

The companies that can best adapt to the changing landscape and leverage their unique skills through strategic partnerships are likely to emerge as the leaders in the SDV era.

The outcome will also depend on consumer preferences. Will drivers prioritize the brand legacy and engineering expertise of traditional automakers, or will the lure of innovative software features and user experiences offered by tech companies be more enticing? Only time will tell who will ultimately control the future of software-defined vehicles.

EIGHTEEN

THE RISE OF SUBSCRIPTION-BASED SERVICES FOR SDVS: A NEW GEAR FOR THE SDV REVOLUTION

The automotive industry is undergoing a seismic shift with the emergence of software-defined vehicles (SDVs). These vehicles prioritize software as the core, enabling continuous improvement, feature upgrades, and customization – all through over-the-air updates. This technological leapfrog presents a compelling opportunity for a new car ownership model: subscription-based services.

Traditional Ownership: A Model on the Brink of Change

Traditionally, car ownership has involved a significant upfront investment followed by ongoing maintenance costs. This model can be financially burdensome, especially for younger generations who prioritize flexibility and access over possession. Additionally, rapid technological advancement renders vehicles obsolete within a few years, leaving owners with outdated technology.

Subscription Services: Geared Towards Flexibility and Access

Subscription-based services for SDVs offer a compelling alternative. Similar to smartphone subscriptions, consumers can choose from various plans offering access to different vehicle types, features, and usage durations. This provides unmatched flexibility as users can switch between car models or features based on their evolving needs.

Furthermore, subscription services take care of maintenance, insurance, and roadside assistance, offering a hassle-free driving experience. Software updates ensure that users always have access to the latest features and functionalities, eliminating the issue of owning a technologically outdated vehicle.

The Rise of the SDV Subscription Model: A Perfect Match

The unique characteristics of SDVs make them particularly well-suited for subscription models. Here's why:

Software-Defined Features: With SDVs, features can be added or removed through software updates. This allows subscription services to offer tiered plans with varying feature sets, catering to diverse customer needs.

Reduced Upfront Costs: Subscription services eliminate the substantial upfront investment associated with traditional car ownership, making access to mobility more affordable for a wider range of consumers.

Seamless Upgrades: Software updates ensure vehicles are always up-to-date with the latest technology, a significant benefit for subscription services that can offer the newest features to their users.

A Transformation in Progress

Subscription services for SDVs are still in their nascent stage, but several car manufacturers are already offering them, paving the way for a future where car ownership becomes more flexible and accessible. Here are some leading examples:

Porsche Drive: Porsche's subscription program allows users to choose from a variety of high-performance models with flexible terms and comprehensive coverage.

Volvo Care by Volvo: This program offers a simplified car ownership experience with a single monthly payment that covers the vehicle, insurance, and maintenance.

General Motors Maven: General Motors' subscription service provides access to a range of GM vehicles, including electric cars, SUVs, and trucks.

BMW Access by BMW: This subscription program allows users to enjoy the latest BMW vehicles with the flexibility to switch vehicles throughout their subscription term.

Lexus Personalized Access: Lexus offers a subscription service with a focus on luxury and convenience, providing users with a curated selection of Lexus vehicles.

Challenges and Considerations

While the potential for subscription services is undeniable, some challenges need to be addressed:

Establishing Pricing Models: Developing fair and competitive pricing models that consider usage, vehicle type, and features will be crucial for attracting customers.

Infrastructure and Logistics: Subscription services require robust infrastructure to manage vehicle fleets, handle maintenance needs, and ensure seamless transitions between users.

Consumer Adoption: Shifting consumer mindsets from ownership to access could be a hurdle. Subscription services need to effectively communicate the benefits and convenience they offer.

The Road Ahead: A Subscription-Based Future?

The rise of SDVs and the growing popularity of the sharing economy have created a fertile ground for subscription-based car ownership. Offering flexibility, affordability, and access to the latest technology, subscription services have the potential to disrupt traditional car ownership models. As technology evolves and consumer preferences shift, a future dominated by subscription-based services for SDVs seems increasingly likely. The companies that can adapt their offerings and effectively cater to the changing needs of consumers will be the ones leading the charge in this exciting new era of mobility.

NINETEEN

The Road Ahead: A Snapshot in Time

So, here we are—at the final chapters of this book, and let me tell you, if you were expecting a grand, all-encompassing conclusion, I hate to break it to you—this book is anything but complete.

Not because I got lazy (okay, maybe just a little), but because SDVs are a living, breathing, evolving phenomenon. By the time you finish reading this page, a startup somewhere is probably cracking a new breakthrough in vehicle autonomy, a tech giant is deploying a more powerful AI-driven cockpit, and an automaker is battling a software bug that wasn't even a concept five years ago.

This book, in its entirety, is just a snapshot in the time scale of SDVs—a Polaroid of where we are today. But if there's one thing I can say with certainty, it's that the success of SDVs, both as a technology and a business model, will revolve around ten fundamental elements. And because I love a good pattern (and because "S" just sounds futuristic), let's call them the 10 S-factors of SDVs.

The 10 S-Factors That Will Define SDVs

1. Software-First Mindset (AI First Mindset)

The car is no longer a mechanical marvel—it's a rolling supercomputer. Companies that treat it as a software-driven ecosystem rather than a hardware-centric product will lead the race. Traditional automakers are still playing catch-up, while tech-first companies are setting the rules of the game.

2. Scalability

Can your SDV software scale from a compact EV to a luxury SUV? Can your business scale across geographies without hitting regulatory roadblocks? Scalability is not just about production—it's about designing a platform that grows, evolves, and adapts with time.

3. Security

Imagine waking up to a notification that your car was hacked while you were sleeping—scary, right? SDVs demand bulletproof cybersecurity, from over-the-air updates to real-time threat detection. In the connected world of tomorrow, security is not a feature—it's a necessity.

4. Seamless Connectivity

An SDV is only as smart as its ability to stay connected—to the cloud, to other vehicles, to infrastructure, and to the driver. 5G, satellite networks, and edge computing will redefine how vehicles interact with their surroundings. Lag is not an option in a world where milliseconds can mean the difference between safety and disaster.

5. Smart Monetization

Cars are no longer just selling points—they're platforms. Subscription-based features, in-car app stores, pay-per-use driving modes... the revenue model of SDVs is shifting from a one-time purchase to a lifetime value ecosystem. The future car dealership might be the app store on your dashboard.

6. Sustainability

SDVs are not just about making cars smarter but making them cleaner. The convergence of software and sustainability will define the next generation of mobility—through energy-efficient software, predictive battery management, and eco-driving algorithms that optimize every mile.

7. Standardization

The industry needs a common language—from software stacks to cybersecurity protocols. Right now, every OEM is building its own version of SDVs, leading to fragmentation. The companies that push for standardization will not only accelerate adoption but also open up new revenue streams through software licensing and ecosystem partnerships.

8. Silicon & Compute Power

Without high-performance computing (HPC) and custom silicon, SDVs are just concept cars in a PowerPoint presentation. The battle for SDVs is happening at the chip level—companies investing in AI accelerators, domain controllers, and power-efficient compute platforms will dominate.

9. Simplicity in User Experience

No one wants to take a two-hour tutorial just to turn on their car's air conditioning. The SDV revolution will succeed only if technology feels intuitive, invisible, and effortless. Think of how Tesla removed most physical buttons in favor of a touchscreen—now imagine an AI-driven cockpit that anticipates your needs before you even think about them. That's where we're heading.

10. Speed of Innovation

The SDV industry is running a marathon at the pace of a sprint. The winners will be those who out-innovate, out-iterate, and outmaneuver the competition. In this space, fast doesn't just win—it survives.

Final Thoughts: A Journey Without a Finish Line

So, there you have it—the 10 S-factors that will define the SDV business model. Will they evolve? Absolutely. Will we see new disruptions that rewrite the rules? Without a doubt.

The reality is that this book is not a final word on SDVs—it's a conversation starter. A checkpoint in a long and unpredictable journey.

Maybe in five years, someone will pick up this book and chuckle at how naïve we were to think SDVs would evolve in a linear path. Maybe in ten years, we'll be debating the rise of AI-driven autonomous taxis that make human driving obsolete.

Whatever the future holds, one thing is certain: the transition from hardware to code is irreversible.

So buckle up—the road ahead is just getting started.

TWENTY

CASE STUDIES

Why Case Studies Matter in 2025—and Beyond

In writing Hardware to Code, I felt it was essential to ground the evolving narrative of Software-Defined Vehicles (SDVs) in the real-world actions of companies that are shaping this transformation. That's why I've included a series of case studies throughout the book—snapshots of how various OEMs, Tier-1 suppliers, startups, and tech firms are navigating the shift from hardware-centric to software-driven mobility.

These stories are not just examples—they are windows into the practical realities, strategic choices, and engineering challenges that define this era of automotive evolution.

Some of the companies featured here may rise to lead the future of mobility; others may pivot, fade, or be disrupted themselves. But that's precisely the point. These case studies capture a moment in time—2025—when the industry was in flux, experimentation, and bold reinvention. They reflect ambition, vision, and risk-taking, and they provide context to understand the broader journey of SDVs.

Years from now, when the dust settles and the winners and losers are clearer, these stories will still hold relevance—not just for who succeeded, but for how the industry thought, adapted, and attempted to innovate at one of its most transformative junctures.

In that spirit, I hope these cases help spark ideas, challenge assumptions, and inspire deeper reflection on your own SDV journey—whether you're an engineer, executive, policymaker, or student of the mobility future.

Why Nio is Poised to Become an Impeccable Brand in China

Nio, formerly known as NEXTEV, has emerged as a serious contender in the ever-evolving Chinese electric vehicle (EV) market. Founded in 2014, the company boasts a unique combination of innovative technology, strategic partnerships, and a deep understanding of the Chinese consumer, making it a prime candidate for impeccable brand status.

A Winning Formula: Technology and Vision

Nio's approach to EVs goes beyond simply manufacturing cars. They have established themselves as a technology leader with several key strengths:

Battery Swap Technology: One of Nio's most significant innovations is its battery swap station network. This allows drivers to quickly exchange depleted batteries for fully charged ones, significantly reducing charging time – a major pain point for EV owners.

Operating System (NIO OS) and Software-Defined Approach: Nio's cars are built on a software-defined platform, allowing for continuous improvement and over-the-air updates. The NIO OS provides a user-centric experience that integrates seamlessly with various in-car features and connectivity options.

Lightweight Design and Performance: Nio vehicles prioritize lightweight materials like aluminum and carbon fiber, resulting in extended range and superior handling. This focus on performance sets them apart from some bulkier EV offerings.

Strategic Partnerships and Brand Building

Nio has strategically partnered with leading companies to bolster its position:

Formula E: Nio's involvement in Formula E, the electric racing championship, serves as a technological proving ground and a powerful marketing tool, showcasing their commitment to innovation and performance.

Investors: Backed by prominent investors like Tencent and Sequoia Capital, Nio has access to significant financial resources to fuel its research and development (R&D) efforts.

Abu Dhabi Government Injects $2.2 Billion into NIO

The deal involves CYVN, an investment firm primarily owned by the Abu Dhabi government, investing a whopping $2.2 billion in NIO.

Beyond the Car: A Lifestyle Experience

Nio understands that in China, car ownership is often interwoven with lifestyle aspirations. They cater to this by offering:

NIO Life: A premium lifestyle brand that extends beyond the vehicle, offering a curated selection of apparel, accessories, and experiences that cater to the Nio customer.

NIO Houses: These community hubs provide Nio owners with a space to connect, relax, and experience the Nio brand beyond just driving.

NIO is carving a unique niche in China's booming electric vehicle (EV) market by focusing on premium SUVs with a clever battery strategy that addresses key consumer concerns.

Luxury and Innovation on the Road

NIO's flagship ES8 SUV, launched in 2019, exemplifies its commitment to luxury and cutting-edge technology. Equipped with advanced driver-assistance systems (ADAS) and an AI system (NOMI), the ES8 competes favorably with established players like Tesla and Audi. Building on the success of the ES8, NIO has expanded its offerings with the ES6, EC6, and most recently, the ET7 sedan. This strategic product portfolio has propelled NIO to the top spot in China's all-electric SUV market, surpassing even Tesla in April 2024.

Conquering Range Anxiety: A Differentiating Powerhouse

NIO's game-changing battery strategy stands out as a major differentiator. They not only offer compatibility with conventional charging stations but also boast a network of battery swap stations across China, with plans for European expansion. This innovative approach eliminates "range anxiety," a significant barrier to widespread EV adoption. Additionally, NIO's Battery-as-a-Service (BaaS) option reduces upfront costs by separating battery purchase from the vehicle itself. By addressing both range and affordability concerns, NIO's battery strategy positions them for long-term success.

Building a Premium Brand Identity

NIO has meticulously crafted a premium brand image. The introduction of the high-performance EP9 in 2016 created an aura of exclusivity and quality, attracting investors and establishing NIO as a serious player in the luxury car market. This strategy effectively countered the perception of Chinese brands lacking in quality, a challenge similar to what Japanese automakers faced after World War II. NIO, like its Japanese counterparts, is leading the charge in changing this perception and showcasing the potential of Chinese innovation on the global stage.

Understanding the Chinese Consumer

Nio has a clear advantage by being a homegrown Chinese company. They are attuned to the specific needs and preferences of Chinese consumers, including:

Range Anxiety: The rapid battery swap network addresses a major concern for Chinese EV buyers.

Government Incentives: China's strong push towards EVs creates a favorable market environment for Nio.

Focus on Technology and Design: Chinese consumers are increasingly tech-savvy and appreciate well-designed products, both areas where Nio excels.

Challenges and the Road Ahead

Despite its strengths, Nio faces some challenges:

Competition: The Chinese EV market is fiercely competitive, with established automakers and well-funded startups vying for market share.

Profitability: As a young company, Nio is not yet profitable and relies on external funding.

Scaling Up: Successfully scaling up production and expanding its service network will be crucial for long-term success.

Conclusion

Nio's innovative technology, strategic partnerships, and deep understanding of the Chinese consumer position it for exceptional brand status. Their commitment to battery swap technology, software-defined vehicles, and a premium lifestyle experience sets them apart in the crowded market. While challenges remain, Nio's unwavering focus on innovation and customer satisfaction makes it a force to be reckoned with in China's electric vehicle revolution.

Rivian: From Wall Street Darling to Electric Truck Enigma – Can It Rev Up Again?

Rivian burst onto the scene with a bang. The electric vehicle (EV) startup, lauded for its sleek trucks and innovative software-defined vehicle (SDV) approach, was once a Wall Street darling. However, its share price has plummeted in recent months, raising concerns about its ability to deliver on its ambitious promises. Let's delve into the challenges Rivian faces and explore potential solutions for a successful comeback.

Rivian, a growing EV maker with innovative vehicles, faces challenges despite its $12.9 billion market cap. While it invests heavily and boasts unique products, ongoing losses and a need for more funding raise concerns. Rivian's success hinges on attracting customers, controlling costs, and navigating a competitive market.

The Thorny Path of a Startup

Several factors are contributing to Rivian's current struggles:

Production Delays: Rivian has faced significant production delays for its R1T pickup truck and R1S SUV. Meeting ambitious production targets is crucial for any automaker, and these delays have eroded investor confidence.

Supply Chain Snags: The global chip shortage and other supply chain disruptions have impacted Rivian's ability to source critical components. While not unique to Rivian, these issues have hampered their progress.

Competition Heats Up: The EV market is becoming increasingly crowded, with established automakers like Ford and General Motors entering the fray. Rivian needs to find a way to stand out in a more competitive landscape.

Burning Cash: As a startup, Rivian is burning through cash to ramp up production. This can be concerning for investors, especially with production delays.

Software-Defined Savior or Achilles' Heel?

Rivian's focus on SDVs – vehicles where software plays a critical role in functionality and performance – is both a potential advantage and a challenge.

Advantage: A robust software platform can offer features like over-the-air updates, improved vehicle performance, and potentially even autonomous driving capabilities. This could be a differentiator in the long run.

Challenge: Developing and integrating complex software requires time and resources. Delays and bugs in the software could further slow down production and damage Rivian's reputation for innovation.

Can Rivian Rev Up Again?

Despite the challenges, Rivian has the potential to turn things around:

Focus on Production: Scaling up production efficiently and meeting deadlines will be crucial to regain investor confidence.

Supply Chain Solutions: Rivian needs to diversify its supply chain and explore alternative sources for critical components to mitigate future disruptions.

Innovation Edge: Continued investment in software development and unique features can help Rivian carve out a niche in the crowded EV market.

Strategic Partnerships: Collaboration with established automakers or technology companies could provide Rivian with access to resources and expertise to accelerate production and development.

While Rivian itself isn't launching entirely new brands, they do have plans for new product lines and market expansions that could act as a form of revival for their future:

The R2 Platform: This is a crucial element in Rivian's growth strategy. The R2 platform is designed to be more affordable and accessible than the R1T and R1S, potentially targeting a wider consumer segment. This could significantly increase Rivian's market share and revenue.

Market Expansion: Rivian isn't just focusing on the US market. Their plans for international expansion, particularly into Europe and China, present a significant growth opportunity. Entering new markets with established EV demand could breathe new life into Rivian's sales figures.

Commercial Vehicles: Rivian isn't just about consumer vehicles. Their commercial vehicle platforms, like the Amazon delivery vans, offer a unique opportunity to tap into a different market segment. Delivering on these contracts and potentially expanding their commercial vehicle offerings could be a major boost for Rivian.

Additionally, Rivian can consider strategic partnerships or collaborations to further its revival:

Co-development with Established Automakers: Partnering with established automakers could provide Rivian with access to resources and expertise, potentially accelerating their production and development processes.

Technology Partnerships: Collaborating with tech companies could enhance Rivian's software-defined vehicle capabilities, potentially giving them a significant edge in the market.

By focusing on these new product lines, market expansions, and strategic partnerships, Rivian can potentially revive its future and establish a strong position in the evolving EV landscape.

The Road Ahead

Rivian's journey will be closely watched by the EV industry. The company has a unique opportunity to redefine the future of electric trucks with its software-defined approach. However, overcoming production hurdles, navigating a competitive landscape, and effectively managing its cash flow will be critical to its success. Only time will tell if Rivian can recapture its "Wall Street darling" status, but the potential for a comeback remains strong.

BYD: A Disruptive Force from the East

Founded in 1995 by Wang Chuanfu, BYD began with a focus on rechargeable batteries. This early commitment to electrification laid the groundwork for their dominance in the SDV space. BYD not only manufactures electric vehicles (EVs), but also the lithium-ion batteries that power them, giving them unmatched control over a critical component of the SDV.

Beyond Batteries: A Feature-Packed Powerhouse

BYD's SDVs boast cutting-edge features that are redefining the driving experience:

Blade Battery Technology: BYD's proprietary Blade Battery offers superior safety, range, and charging speeds compared to traditional lithium-ion batteries.

DiLink Intelligent Network System: A user-friendly, customizable in-car infotainment system with voice control, facial recognition, and seamless smartphone integration.

Driver-Assistance Systems: Advanced ADAS features like automatic emergency braking, lane departure warning, and adaptive cruise control enhance safety and comfort.

Over-the-Air (OTA) Updates: BYD vehicles receive regular software updates, ensuring they stay at the forefront of technology and receive new features throughout their lifespan.

From Humble Beginnings to Global Aspirations:

BYD's success story is as inspiring as it is disruptive. From its roots in battery production, the company has transformed into a leading EV manufacturer, not just in China, but globally. Their strategic overseas investments and partnerships position them for continued growth. Notably, BYD is a major player in the South American EV market, showcasing their ambition beyond China.

A Status Symbol Redefined:

Owning a BYD is no longer just about practicality. Their sleek designs, cutting-edge technology, and focus on sustainability have made them desirable and aspirational. BYD is changing the perception of Chinese cars, proving they can be at the forefront of automotive innovation.

Why This Matters:

BYD's rise to prominence in the SDV space signifies a significant shift in the global automotive landscape. Chinese companies are no longer simply competitors, but leaders in pushing the boundaries of technology and design. This has major implications for established automakers who will need to adapt and innovate to keep pace.

The Road Ahead

As the SDV revolution continues to unfold, BYD is poised to play a pivotal role. Their commitment to research and development, coupled with their unique battery technology and software expertise, positions them as a force to be reckoned with. With their continued innovation and global aspirations, BYD is not just leading the SDV revolution in China, but shaping the future of mobility worldwide.

BYD's rise in the software-defined vehicle (SDV) space isn't just about batteries; it's about a user-centric approach and a commitment to innovation. Here's a breakdown of some key user and unique features that set BYD cars apart:

User-Friendly Tech for Everyone:

DiLink Intelligent Network System: This in-car infotainment system is the heart of the BYD user experience. It boasts:Voice Control: Keep your eyes on the road with hands-free control for navigation, music, and phone calls.Facial Recognition: Personalized settings and preferences automatically adjust based on the driver. Seamless Smartphone Integration: Effortlessly connect your phone for calls, music streaming, and navigation apps.

Driver Assistance Systems (ADAS): BYD prioritizes safety with features like:Automatic Emergency Braking (AEB): Helps prevent collisions by automatically applying the brakes in critical situations. Lane Departure Warning (LDW): Alerts you if you unintentionally stray from your lane.Adaptive Cruise Control (ACC): Maintains a safe following distance from the car ahead.

Over-the-Air (OTA) Updates: BYD vehicles receive regular software updates, ensuring you always have the latest features and bug fixes. This keeps your car feeling fresh and technologically advanced throughout its lifespan.

Beyond the Basics: Unique BYD Innovations

Blade Battery Technology: This proprietary lithium-ion battery design offers several advantages:Superior Safety: The Blade Battery boasts exceptional thermal stability, minimizing the risk of fire. Increased Range: BYD vehicles achieve impressive driving ranges thanks to the Blade Battery's high energy density.Faster Charging Speeds: Spend less time at charging stations with BYD's efficient battery technology.

360° Panoramic Camera System: Maneuvering tight spaces and parking is a breeze with a complete view of your surroundings.

Phone-as-Key Functionality: Leave your traditional key fob at home. Use your smartphone to unlock, lock, and even start your BYD car.

Who Supplies the Parts? Exploring BYD's Tier 1 Network

While BYD manufactures their Blade Batteries in-house, they collaborate with various Tier 1 suppliers for other vehicle components. Identifying specific suppliers can be challenging due to potential variations across different BYD models and regions. However, some known Tier 1 suppliers for BYD include:

Bosch: A global leader in automotive technology, Bosch likely supplies components like engine control units (ECUs), braking systems, and electronic stability control (ESC) systems.

Continental: Another major automotive supplier, Continental might provide components like tires, instrument clusters, and driver assistance systems.

Denso: This Japanese multinational specializes in automotive components and could be a supplier for BYD's air conditioning systems, powertrain components, and instrument panels.

BYD: A User-Centric Approach to the Future of Mobility

By prioritizing user-friendly technology, safety features, and unique innovations like the Blade Battery, BYD is shaping the future of SDVs. Their focus on in-car connectivity, driver assistance, and over-the-air updates creates a truly connected and personalized driving experience. As BYD continues to collaborate with established Tier 1 suppliers and develop their own technologies, they are poised to remain a leader in the ever-evolving automotive landscape.

SOCs and Operating systems

BYD has not publicly disclosed specific information about the System-on-Chip (SoC) chips or operating systems used in their vehicles. However, given the emphasis on advanced technology and connectivity in their vehicles, it's likely that BYD utilizes SoCs from reputable semiconductor manufacturers known for automotive applications.

Some potential suppliers of SoCs for automotive use include:

Nvidia: Nvidia's DRIVE platform is widely used in autonomous driving systems and offers powerful processing capabilities suitable for advanced driver assistance systems (ADAS) and autonomous driving features.

Qualcomm: Qualcomm's Snapdragon Automotive platform provides scalable solutions for infotainment, connectivity, and telematics systems in vehicles.

Intel: Intel's Atom and Core processors are utilized in automotive applications for in-vehicle infotainment (IVI) systems and other computing tasks.

Renesas Electronics: Renesas offers a range of SoCs tailored for automotive applications, including microcontrollers (MCUs) and system-on-chips (SoCs) for powertrain, safety, and connectivity.

Regarding operating systems, automotive manufacturers typically use proprietary or customized software stacks for their vehicles. It's common for carmakers to develop their own operating systems or adapt existing ones to suit their specific requirements. Some examples of operating systems used in automotive applications include:

Automotive Grade Linux (AGL): AGL is an open-source Linux-based operating system designed for automotive applications. It provides a flexible and customizable platform for infotainment, telematics, and other in-vehicle systems.

QNX: QNX is a real-time operating system (RTOS) commonly used in automotive infotainment and safety-critical systems. It offers reliability, security, and scalability for embedded automotive applications.

Android Automotive: Google's Android Automotive platform is gaining traction in the automotive industry, offering a familiar user experience and integration with Google services for infotainment and connectivity features.

Customized Solutions: Some car manufacturers develop their own proprietary operating systems or software stacks tailored to their specific needs and requirements. These custom solutions may be based on existing platforms or developed from scratch in-house.

Given BYD's focus on innovation and technology, they are likely to leverage advanced SoCs and customized operating systems to power their vehicles' infotainment, connectivity, and autonomous driving features. However, without specific information from

BYD, the exact details of their hardware and software choices remain proprietary.

TESLA

Tesla indeed revolutionized the automotive industry with its software-defined approach, innovative features, and disruptive strategies. Here are 15 key features and strategies that have made Tesla unique:

Electric Vehicle (EV) Technology: Tesla's use of electric propulsion systems instead of traditional internal combustion engines was a game-changer. This technology offered benefits like zero emissions, instant torque, and reduced reliance on fossil fuels, contributing to the EV revolution.

Battery Technology: Tesla developed advanced battery technology, including its proprietary lithium-ion battery cells and battery management systems. This enabled longer range, faster charging times, and improved energy density, setting Tesla apart from its competitors.

Autonomous Driving Capability: Tesla's vehicles are equipped with hardware for autonomous driving, and the company continuously improves its Autopilot software through over-the-air updates. This feature has positioned Tesla as a leader in self-driving technology.

Over-the-Air Updates: Tesla pioneered the use of over-the-air software updates to enhance vehicle functionality, fix bugs, and introduce new features. This agile approach to software deployment allows Tesla to continually improve its vehicles' performance and user experience.

Direct Sales Model: Tesla sells its vehicles directly to consumers through company-owned stores and its website, bypassing traditional dealership networks. This direct sales model allows Tesla to control the customer experience, pricing, and distribution channels.

Vertical Integration: Tesla vertically integrates many aspects of its supply chain, including battery production, vehicle assembly, and software development. This approach gives Tesla greater control over quality, costs, and innovation.

Gigafactories: Tesla's Gigafactories are large-scale manufacturing facilities dedicated to producing batteries, electric drivetrains, and vehicles. These Gigafactories enable Tesla to scale production and reduce manufacturing costs through economies of scale.

Supercharger Network: Tesla operates a global network of Supercharger stations, offering fast-charging capabilities for its vehicles. This infrastructure investment alleviates range anxiety for Tesla owners and promotes the adoption of electric vehicles.

Premium Brand Image: Tesla has cultivated a premium brand image associated with innovation, sustainability, and luxury. The company's sleek designs, cutting-edge technology, and celebrity endorsements contribute to its aspirational appeal.

Elon Musk's Visionary Leadership: Elon Musk, Tesla's CEO, is known for his bold vision, entrepreneurial spirit, and charismatic leadership style. Musk's personal brand and public persona have become closely intertwined with Tesla's identity, shaping its reputation and driving media attention.

Market Disruption: Tesla disrupted the automotive industry by challenging conventional wisdom and established players. The company's disruptive strategies, including direct sales, over-the-air updates, and vertical integration, have forced competitors to adapt and innovate.

Environmental Sustainability: Tesla's focus on electric vehicles aligns with growing concerns about climate change and environmental sustainability. By promoting clean energy transportation, Tesla has positioned itself as a leader in the transition to a low-carbon economy.

Performance and Innovation: Tesla vehicles offer impressive performance metrics, including acceleration, top speed, and handling. The company's relentless focus on innovation has led to groundbreaking features like Ludicrous Mode, bioweapon defense mode, and the Tesla Roadster's space-bound payload.

Customer Engagement: Tesla fosters a passionate community of customers and enthusiasts through events, social media engagement, and owner forums. This grassroots support contributes to brand loyalty and word-of-mouth marketing.

Disruptive Pricing Strategies: While initially targeting the luxury segment with models like the Roadster and Model S, Tesla has progressively introduced more affordable vehicles like the Model 3 and Model Y. The company's pricing strategies, combined with government incentives and cost reductions, have made electric vehicles increasingly accessible to mainstream consumers.

Overall, Tesla's success can be attributed to its combination of innovative technology, disruptive business strategies, strong brand identity, and visionary leadership. As the automotive industry continues to evolve, Tesla remains at the forefront of innovation and transformation.

How to build the next TESLA: A multi-pronged approach

Tesla's disruptive impact on the automotive industry is undeniable. To create a competitor that can not only rival but surpass Tesla, a comprehensive strategy encompassing every stage of car development is necessary. This essay will explore key areas where excellence is crucial: architecture, technology, supply chain, manufacturing quality, chip design, and future-oriented design thinking.

1. Architectural Innovation:

Beyond the Sedan: Move past traditional car designs and explore innovative vehicle architectures. Think electric pickup trucks, futuristic passenger pods, or even flying cars (within safety and regulatory limitations).

Lightweight Materials: Utilize advanced materials like carbon fiber and high-strength aluminum to reduce weight and improve efficiency.

Aerodynamic Optimization: Design a car with a sleek, low-drag coefficient for maximum range and energy savings.

2. Cutting-Edge Technology:

Next-Gen Battery Technology: Invest in solid-state battery research for increased range, faster charging times, and improved safety.

Advanced Driver-Assistance Systems (ADAS): Develop a robust ADAS suite that surpasses Tesla's Autopilot, offering features like self-parking, lane-changing assistance, and potentially true autonomous driving capabilities (regulations permitting).

Connected Car Experience: Create an intuitive and personalized in-car infotainment system with features like voice control, seamless smartphone integration, and over-the-air software updates.

3. Building a Resilient Supply Chain:

Strategic Partnerships: Forge strategic partnerships with key suppliers of critical components like batteries, motors, and semiconductors.

This could involve joint ventures or vertical integration to secure resources and avoid supply chain disruptions.

Sustainable Practices: Implement sustainable sourcing practices throughout the supply chain, minimizing environmental impact and ensuring ethical labor practices.

Global Reach: Establish a robust global supply chain with diverse manufacturing locations to mitigate risks associated with geopolitical instability and resource shortages.

4. Manufacturing Excellence:

Automation and Efficiency: Invest in automation and lean manufacturing principles to streamline production and maintain high quality standards.

Focus on Quality Control: Implement rigorous quality control measures at every stage of production to eliminate defects and ensure consistent performance.

Data-Driven Manufacturing: Utilize data analytics to monitor production processes, identify bottlenecks, and continuously improve efficiency and quality.

5. Designing the Future Chip:

In-House Chip Design: Consider developing custom-designed chips specifically optimized for electric vehicles and autonomous driving functions. This could offer performance advantages over commercially available chips.

Collaboration with Chipmakers: Partner with leading chipmakers to co-develop cutting-edge chips that cater to the specific needs of electric vehicles.

Futureproofing Chip Design: Design chips with scalability and future technological advancements in mind, ensuring continued performance as technology evolves.

6. Visionary Design Thinking:

Sustainable Design: Integrate sustainability into every aspect of car design, from material selection to end-of-life vehicle recycling.

User-Centric Design: Prioritize user experience by designing intuitive interfaces, comfortable interiors, and innovative features that cater to evolving consumer needs.

Adaptability and Upgradability: Design vehicles that can be easily adapted and upgraded with new technologies, ensuring the car stays relevant for an extended lifespan.

Building a car that surpasses Tesla is a monumental task. By focusing on these key areas and fostering a culture of continuous innovation, a new competitor can emerge, pushing the boundaries of electric vehicle technology and design, ultimately shaping a more sustainable and exciting future for transportation.

BMW

Software-Defined Advancements:

BMW has made significant strides in embracing software-defined advancements to enhance the driving experience and differentiate itself in the competitive automotive market. One of the key areas of focus for BMW has been in-car software systems, particularly with their Operating System 7 (OS 7). OS 7 integrates various functions within the vehicle, including navigation, infotainment, connectivity, and vehicle settings, into a cohesive and user-friendly interface. This software-defined approach allows for seamless updates and customization, enabling BMW to adapt to changing consumer preferences and technological advancements rapidly.

Additionally, BMW has leveraged software-defined architecture to introduce advanced driver-assistance systems (ADAS) and autonomous driving features in their vehicles. Through the integration of sensors, cameras, and AI algorithms, BMW vehicles can offer capabilities such as lane-keeping assist, adaptive cruise control, and automated parking, enhancing both safety and convenience for drivers.

Architecture:

BMW's architecture, particularly with OS 7, is designed to be modular and scalable, allowing for integration with a wide range of hardware components and third-party applications. This flexibility enables BMW to keep pace with technological advancements and tailor their offerings to meet the diverse needs of consumers worldwide.

Unique Features Compared to Competitors:

Compared to Japanese and Chinese competitors, BMW's software-defined architecture offers several unique features:

Integration and Customization: BMW's OS 7 allows for seamless integration with smartphones and third-party apps, providing users with a familiar and personalized experience. This level of integration sets BMW apart from some Japanese competitors who may have more closed ecosystems.

Performance and Luxury: BMW's focus on performance and luxury remains a hallmark of its brand identity, even within the realm of software-defined advancements. While Chinese competitors may prioritize affordability and mass-market appeal, BMW continues to emphasize premium features and driving dynamics.

Global Brand Recognition: BMW's strong brand recognition and reputation for engineering excellence give it a competitive edge over both Japanese and Chinese competitors, particularly in markets where prestige and status play a significant role in purchasing decisions.

Future Moves and Technology Partnerships:

Looking ahead, BMW is likely to continue investing in software-defined advancements to stay ahead of the competition and meet evolving consumer expectations. This may involve further collaborations with technology companies to integrate cutting-edge features such as augmented reality navigation, natural language processing, and advanced AI assistants into their vehicles.

One notable technology partnership for BMW is its collaboration with NVIDIA to develop a new generation of in-car computing platforms. By leveraging NVIDIA's expertise in AI and automotive computing, BMW aims to enhance the performance and capabilities of its ADAS and autonomous driving systems, further cementing its position as a leader in automotive innovation.

Operating System Insights:

BMW's Operating System 7 is built on a Linux-based platform, providing a stable and secure foundation for its in-car software systems. The architecture is designed to support over-the-air (OTA) updates, allowing BMW to deliver new features and enhancements to customers remotely. This approach not only ensures that BMW vehicles remain up-to-date with the latest technology but also minimizes the need for costly and time-consuming visits to service centers.

Cloud Partnership Insights:

Amazon Web Services, Inc. (AWS) and the BMW Group have announced a strategic collaboration to develop customizable cloud software aimed at simplifying the distribution and management of data from connected vehicles. This innovative software, which will serve as the foundation for BMW's next-generation, cloud-based vehicle data platform, marks the first time an automaker will utilize such technology. The partnership aims to make this software available to other automakers, enabling them to integrate vehicle data sources, accelerate feature development, and enhance driver experiences at reduced costs.

The jointly developed solution collects vehicle signals and fleet intelligence data, processes it securely in the cloud, and integrates AWS capabilities such as analytics and machine learning. This empowers BMW Group experts to create new vehicle features and applications while ensuring data privacy and compliance with governance policies. AWS's extensive cloud infrastructure and industry-leading security bolster the platform's capabilities, positioning it to unlock next-generation vehicle capabilities and support the automotive industry's evolution towards electric vehicles, software-defined development, digital services, and autonomous driving features.

The BMW Group will benefit from this software as part of AWS's offering for the automotive industry, enabling add-on features for vehicles and responsive applications for fleet management. The platform facilitates centralized computing, continuous software development, real-time data management, and machine learning, laying the groundwork for future mobility solutions, including autonomous and electric vehicles.

Survival Amidst Competition from EVs:

As the automotive industry continues to shift towards electric vehicles (EVs), BMW faces challenges in adapting its traditional combustion-engine lineup to meet stricter emissions regulations and changing consumer preferences. However, BMW has already made significant investments in electric mobility, with the introduction of models such as the BMW iX3 and BMW i4.

To survive and thrive amidst competition from EVs, BMW is likely to continue expanding its electric vehicle portfolio, investing in battery technology, and enhancing its charging infrastructure. Additionally, BMW's focus on software-defined advancements will enable it to differentiate its EV offerings through innovative features and user experiences, helping to maintain its competitive edge in the evolving automotive landscape.

NVIDIA

The automotive industry is undergoing a transformative shift towards software-defined vehicles (SDVs), where software takes a central role in controlling and enhancing vehicle functions. NVIDIA, a leader in graphics processing and artificial intelligence (AI), is at the forefront of this revolution. This case study explores how NVIDIA's innovations and platforms are driving the development of SDVs, providing a glimpse into the future of automotive technology.

Background

NVIDIA, traditionally known for its powerful GPUs used in gaming and professional graphics, has expanded its expertise into various fields, including artificial intelligence, data science, and autonomous driving. Leveraging its strengths in AI and high-performance computing, NVIDIA has developed comprehensive solutions for the automotive industry aimed at enabling software-defined capabilities in vehicles.

NVIDIA DRIVE Platform

The cornerstone of NVIDIA's strategy in the automotive sector is the NVIDIA DRIVE platform, an end-to-end solution for autonomous and semi-autonomous vehicles. The platform integrates hardware and software to provide a scalable and flexible environment for developing, testing, and deploying AI-driven vehicle functions.

Key Components of the NVIDIA DRIVE Platform:

DRIVE AGX: A scalable computing platform designed to handle the high computational demands of autonomous driving. It integrates NVIDIA's Xavier and Orin SoCs (System on Chips), offering powerful processing capabilities for AI and sensor data processing.

DRIVE OS:The foundational operating system that supports the development of automotive applications. DRIVE OS includes the necessary libraries, APIs, and tools to leverage NVIDIA's hardware effectively.

DRIVE AV and DRIVE IX:DRIVE AV (Autonomous Vehicle) focuses on enabling full autonomous driving capabilities through advanced perception, planning, and control algorithms. DRIVE IX (Intelligent Experience) enhances in-car experiences with AI-driven features, such as advanced driver assistance, in-cabin monitoring, and personalized user interactions.

DRIVE Hyperion:An open and scalable hardware platform that provides a complete sensor suite for autonomous driving, including cameras, radar, lidar, and ultrasonics, integrated with NVIDIA's compute platform.

Case Study: Volvo and NVIDIA

One of the prominent partnerships showcasing NVIDIA's impact on SDVs is with Volvo. In 2020, Volvo announced that it would leverage NVIDIA's DRIVE Orin platform to power its next-generation vehicles.

Objectives:

To develop a robust and scalable platform for autonomous driving and advanced driver assistance systems (ADAS).

To enhance in-car experiences with intelligent features driven by AI.

Implementation:

Volvo integrated NVIDIA DRIVE Orin into its vehicle architecture, enabling high-performance computing necessary for processing vast amounts of sensor data in real-time.

The partnership focused on software-first development, allowing Volvo to continuously update and improve vehicle functionalities through over-the-air (OTA) updates.

Utilizing NVIDIA's DRIVE AGX and Hyperion platforms, Volvo could develop and test new autonomous driving capabilities in a safe and controlled environment.

Outcomes:

Enhanced safety and driving assistance features, contributing to Volvo's vision of zero fatalities in its vehicles.

Improved user experience through AI-powered features like voice recognition, driver monitoring, and personalized settings.

A future-proof vehicle architecture capable of adapting to new technologies and consumer demands through continuous software updates.

Challenges and Solutions

Challenge: High Computational Requirements

Autonomous driving requires immense computational power to process data from multiple sensors and execute complex AI algorithms in real-time.

Solution: Scalable Computing with DRIVE AGX

NVIDIA's DRIVE AGX platform, with its powerful GPUs and AI accelerators, provides the necessary computational resources. Its scalable architecture allows automakers to customize their solutions according to specific requirements.

Challenge: Ensuring Safety and Reliability

Autonomous and semi-autonomous systems must operate with a high degree of safety and reliability to gain consumer trust and meet regulatory standards.

Solution: Comprehensive Testing and Validation

NVIDIA DRIVE includes tools for simulation and validation, such as DRIVE Sim, which allows automakers to test their systems in a virtual environment. This ensures rigorous testing under diverse scenarios, enhancing safety and reliability.

Future Prospects

NVIDIA continues to push the boundaries of what is possible with SDVs. Future developments include more advanced AI capabilities, enhanced connectivity features, and deeper integration with smart city infrastructures. NVIDIA's focus on software-defined platforms ensures that vehicles can continuously evolve, offering new features and improvements long after they leave the showroom floor.

Conclusion

NVIDIA's contributions to the development of software-defined vehicles highlight the transformative potential of integrating advanced computing and AI into the automotive industry. Through strategic partnerships, innovative platforms, and a commitment to continuous improvement, NVIDIA is driving the evolution of transportation towards a smarter, safer, and more dynamic future. The case of Volvo exemplifies how traditional automakers can leverage NVIDIA's technology to stay ahead in the rapidly evolving landscape of automotive technology.

INTEL

Intel: Powering the Software-Defined Vehicle Revolution

The automotive industry is undergoing a transformative shift towards software-defined vehicles (SDVs). These vehicles prioritize software as the core, enabling continuous improvement, feature upgrades, and customization – all through over-the-the-air updates. In this dynamic landscape, Intel plays a crucial role in shaping the future of SDVs, leveraging its expertise in hardware, software, and connectivity solutions.

A Legacy of Innovation: From Hardware to Software

For decades, Intel has been a dominant force in the automotive industry, supplying essential hardware components like microprocessors and controllers. This established foundation in automotive electronics provides Intel with a deep understanding of the unique needs and challenges of the car manufacturing process.

However, Intel is not resting on its laurels. Recognizing the software-centric future of vehicles, the company is actively expanding its role in the SDV ecosystem. Here's a glimpse into some of their recent advancements, echoing the themes discussed at CES 2024:

Leading the Charge in Open Standards: At CES 2024, Intel highlighted its commitment to open automotive chiplet platforms. This initiative, as presented by Jack Weast, VP and GM of Intel Automotive, aims to create a standardized approach for integrating various processing units within a single chip. This fosters greater flexibility and scalability for automakers when developing SDVs.

Pioneering Power Management Solutions: Another key takeaway from CES 2024 was Intel's focus on vehicle power management. Their announcement of being the first supplier to deliver an automotive-grade chiplet-based product with industry leader, MACOM, signifies a significant step towards optimizing energy consumption in SDVs. This focus on efficient power management aligns perfectly with the growing need for extended battery life in electric vehicles, a major area of concern for the industry.

Mobileye: A Strategic Advantage: Intel's acquisition of Mobileye, a leader in computer vision and driver-assistance systems, continues to be a strategic advantage. Mobileye's technology forms the backbone of Intel's self-driving car solutions, as mentioned earlier.

The Impact on SDVs: From Safety to User Experience

Intel's contributions are shaping SDVs in several crucial ways:

Enhanced Safety Features: Intel's technology helps develop advanced driver-assistance systems (ADAS) and autonomous driving functionalities, ultimately aiming to reduce accidents and improve road safety.

Seamless Over-The-Air Updates: Intel's hardware and software solutions enable secure and efficient over-the-air (OTA) updates, allowing carmakers to deliver new features and bug fixes to vehicles remotely.

Personalized User Experiences: Intel's technology paves the way for in-vehicle AI that can personalize the driving experience, from adjusting cabin temperature to recommending music based on driver preferences.

Looking Forward: The Road Ahead

As the SDV revolution continues, Intel remains a key player at the forefront. With its commitment to innovation, open collaboration, and a diverse portfolio of hardware and software solutions, Intel is well-positioned to shape the future of mobility. By continually pushing the boundaries of what's possible, Intel aims to ensure that SDVs are safe, reliable, and deliver an exceptional user experience for drivers and passengers alike.

The road ahead promises exciting developments in the world of SDVs, and Intel, as evidenced by their CES 2024 announcements, is undoubtedly a company to watch as it continues to play a pivotal role in this transformative era.

QUALCOMM

The automotive industry is undergoing a seismic shift towards software-defined vehicles (SDVs). These vehicles prioritize software as the core, enabling features like over-the-air updates, continuous improvement, and customization. To accelerate this transformation, Amazon Web Services (AWS) and Qualcomm Technologies Inc. have joined forces to create a demonstrator for cloud-native development using the Snapdragon Digital Chassis platform.

What is the Snapdragon Digital Chassis?

Qualcomm's Snapdragon Digital Chassis is a comprehensive platform designed to provide a foundation for the development of next-generation vehicles. It integrates various processing units, software frameworks, and development tools, enabling automakers to develop feature-rich and highly connected vehicles.

The Role of AWS in Cloud-Native Development

Cloud-native development is an approach where applications are designed specifically to run in the cloud environment. This approach offers several benefits for SDV development, including:

Scalability: Cloud resources can be easily scaled up or down to meet the demands of development and deployment.

Agility: Cloud-based development allows for faster iteration and testing cycles.

Cost-efficiency: Automakers only pay for the resources they use.

The AWS and Qualcomm demonstrator leverages these advantages by providing developers with access to:

High-performance computing resources: AWS Graviton processors and Amazon DL2q instances powered by Qualcomm Cloud AI 100 offer the necessary processing power for demanding workloads like vehicle simulation and software testing.

Snapdragon Virtual Platforms: These platforms enable developers to run the same base platform software used in actual vehicles on the AWS cloud. This provides a realistic development environment without the need for physical hardware.

Development and deployment tools: AWS offers a wide range of tools and services for developing, testing, and deploying cloud-native applications, streamlining the entire SDV development process.

Benefits of the Collaboration

The collaboration between AWS and Qualcomm offers several benefits for the automotive industry:

Faster Development Cycles: The cloud-native approach facilitates quicker development and deployment cycles for SDV features and functionalities.

Reduced Costs: By leveraging the scalability and cost-efficiency of the cloud, automakers can potentially reduce their development and deployment costs.

Improved Innovation: The demonstrator provides developers with a powerful platform to experiment with new ideas and accelerate innovation in the SDV space.

Looking Forward: A Collaborative Future

The partnership between AWS and Qualcomm represents a significant step forward for the development of SDVs. By leveraging the strengths of cloud computing and the Snapdragon Digital Chassis platform, this collaboration has the potential to revolutionize the automotive industry. As cloud-native development becomes the norm for SDVs, similar collaborations between technology giants and automotive players will likely become increasingly common, shaping the future of mobility.

INFINEON

Infineon: Powering the Software-Defined Vehicle Revolution

The automotive industry is undergoing a metamorphosis, with software taking center stage. This shift towards software-defined vehicles (SDVs) presents both challenges and opportunities for established players like Infineon Technologies. This essay will explore Infineon's product portfolio, services, and solutions specifically designed for SDVs, along with their strategic vision for this evolving landscape.

Infineon's Software-Defined Vehicle Arsenal

Infineon is not a newcomer to the automotive scene. For decades, their high-performance semiconductors have been instrumental in vehicle electronics. However, recognizing the software-centric future, Infineon has significantly expanded its offerings:

Microcontrollers: Infineon's AURIX™ family of microcontrollers provides the robust hardware foundation for SDVs. These safety-certified microcontrollers are crucial for handling complex software tasks and ensuring reliable vehicle operation.

Connectivity Solutions: Secure and reliable connectivity is vital for SDVs, allowing for over-the-air (OTA) updates, data exchange with cloud platforms, and vehicle-to-everything (V2X) communication. Infineon's comprehensive portfolio of connectivity solutions, including cellular modules and security chips, addresses these needs.

Power Electronics: The rise of electric vehicles (EVs) within the SDV landscape necessitates efficient power management. Infineon's expertise in power semiconductors, like CoolMOS™ power MOSFETs and Econo™ inverters, ensures optimal power conversion and system efficiency for EVs.

Software Development Tools: Infineon recognizes that hardware is only half the equation. They offer software development tools and platforms like AURIX™ Toolbox, which simplifies software development for complex automotive applications.

Partnerships: Infineon understands that the SDV revolution requires collaboration. They actively partner with companies like Green Hills Software and Apex.AI to develop integrated hardware-software platforms that streamline development and accelerate innovation.

Infineon's Vision for SDVs: A Symphony of Hardware and Software

Infineon views SDVs not just as a technological shift but as a paradigm change. They believe that the true potential of SDVs can only be realized by fostering a harmonious relationship between cutting-edge hardware and powerful software. Here's how their strategy reflects this vision:

Focus on Safety and Security: Safety remains paramount in the automotive industry. Infineon prioritizes safety-certified hardware and software solutions to ensure the integrity and reliability of SDV functionalities.

Openness and Collaboration: Infineon champions open standards and actively collaborates with other industry players. This fosters innovation and creates a more robust ecosystem for SDV development.

Evolving with Technology: The world of SDVs is constantly evolving. Infineon is committed to continuous innovation, ensuring their solutions stay at the forefront of technological advancements.

Looking Forward: Steering the Course of Innovation

Infineon's comprehensive product portfolio, commitment to partnerships, and focus on safety and security position them as a key player in the SDV revolution. As the industry navigates this software-defined future, Infineon is actively shaping the course of innovation by offering the essential building blocks for the next generation of vehicles. By fostering a collaborative environment and continuously evolving its technology offerings, Infineon aims to be a driving force in creating a safer, more connected, and ultimately, more enjoyable driving experience for everyone.

BOSCH

Bosch: Charting the Course for Software-Defined Vehicles

Bosch, a long-standing leader in automotive technology, is at the forefront of the software-defined vehicle (SDV) revolution. Their comprehensive approach to SDVs encompasses not just hardware solutions but also a robust software architecture, vehicle service-oriented architecture (SOA), and a vision for a future with in-vehicle app stores. Let's delve deeper into Bosch's approach to SDVs:

The Bosch SDV Philosophy: Beyond Hardware

While Bosch is a powerhouse in automotive components like sensors and control units, they recognize that software is the future. Their SDV approach goes beyond mere hardware by focusing on:

Software-Defined Functions: Bosch develops software-defined functions (SDFs) that control various vehicle aspects, from powertrain management to advanced driver-assistance systems (ADAS). This modular approach allows for flexibility and continuous improvement through over-the-air (OTA) updates.

Openness and Standardization: Bosch understands that a closed ecosystem stifles innovation. They promote open standards and actively collaborate with other industry players. This fosters a collaborative environment where diverse expertise can contribute to the advancement of SDVs.

Security by Design: Cybersecurity is a top priority in the software-driven world. Bosch integrates security features throughout the development process, ensuring the integrity of software and protecting vehicles from cyber threats.

The Bosch Software Architecture: Building the Foundation

The foundation of any successful SDV is a robust software architecture. Bosch's architecture comprises several key elements:

Basic Software (BSW): This layer provides a standardized platform for developing applications. It includes essential functionalities like communication protocols, operating systems, and device drivers.

Middleware: This layer acts as a bridge between the BSW and the application software. It manages communication between different software components and ensures smooth data exchange.

Application Software: This layer houses the actual SDFs that control various vehicle features and functionalities. These applications can be developed by Bosch or third-party companies using standardized APIs.

Vehicle SOA: Orchestrating the Symphony

The vehicle SOA is a crucial concept in Bosch's SDV approach. It allows different software components to communicate and exchange data seamlessly. Imagine an orchestra where various instruments (software components) work together to create a harmonious symphony (vehicle functionality). The SOA acts as the conductor, ensuring all components play their part in a coordinated manner.

Layers of the Vehicle SOA:
The Bosch vehicle SOA can be further broken down into layers:

Service Layer: This layer defines the services offered by different software components.

Communication Layer: This layer provides the communication protocols for exchanging data between services.

Data Layer: This layer manages the data used by the services, ensuring its integrity and security.

Vehicle App Store: The Future of In-Vehicle Customization

Bosch envisions a future where drivers can personalize their driving experience through a vehicle app store. Similar to smartphone app stores, this platform would allow users to download and install new features and functionalities for their vehicles. This could include anything from advanced navigation apps to entertainment applications or even personalized driving profiles.

The Road Ahead: Bosch at the Wheel

Bosch's commitment to open standards, robust software architecture, and a collaborative vision positions them as a leader in the SDV revolution. Their approach, encompassing hardware, software, and a vision for future in-vehicle customization, paves the way for a future where vehicles are not just machines, but intelligent and adaptable companions on the road. As the industry navigates this software-defined future, Bosch is at the wheel, steering the course towards a safer, more connected, and ultimately more personalized driving experience for everyone.

NXP SEMICON

NXP: Building the Blocks for the Software-Defined Vehicle Revolution

The automotive industry is undergoing a transformative shift towards software-defined vehicles (SDVs). These vehicles prioritize software as the core, enabling features like over-the-air updates, continuous improvement, and customization. In this dynamic landscape, NXP Semiconductors is playing a crucial role by providing the essential building blocks for the next generation of vehicles.

NXP's Three Pillars of Support for SDVs

NXP recognizes that SDV development requires a holistic approach. They offer a comprehensive portfolio of solutions, focusing on three key pillars:

1. The Silicon: S32 Automotive Platform

At the heart of NXP's offering lies the powerful S32 Automotive Platform. This family of microprocessors and microcontrollers provides the robust hardware foundation for SDVs.

Processing Power: S32 processors deliver the necessary processing power to handle complex software tasks, including real-time data analysis and advanced driver-assistance systems (ADAS) functionalities.

Scalability: The S32 platform offers a range of processors with varying levels of processing power and memory, allowing automakers to choose the best fit for their specific needs.

Functional Safety: NXP prioritizes safety in its automotive products. The S32 platform includes features like hardware redundancy and error correction mechanisms, ensuring the reliable operation of critical vehicle functions.

2. The Analog Components: In-Vehicle Networking and Power Management

Beyond processors, NXP offers a wide range of analog components that are essential for efficient and reliable SDV operation:

In-Vehicle Networking: High-speed networking solutions like Ethernet enable seamless communication between various electronic control units (ECUs) within the vehicle. This allows for real-time data exchange and coordinated operation of different vehicle systems.

Power Management: Efficient power management is crucial for modern vehicles, especially electric vehicles (EVs). NXP's power management solutions ensure optimal power conversion and system efficiency, maximizing battery life and vehicle range.

3. The Ecosystem: Collaboration and Partnership

NXP understands that no single company can build the future of mobility alone. They actively foster collaboration within the automotive ecosystem by:

Open Standards: NXP supports open standards for software and hardware interfaces. This facilitates easier integration of components from different vendors and promotes a more flexible development environment.

Partnering with Industry Leaders: NXP collaborates with leading software companies, automotive manufacturers, and other industry players to create a comprehensive ecosystem that supports the development of next-generation SDVs.

The Benefits for OEMs

NXP's three-pronged approach offers significant advantages for original equipment manufacturers (OEMs) developing SDVs:

Reduced Development Time and Cost: The S32 platform's scalability and pre-integrated features can streamline the development process, leading to faster time-to-market and reduced development costs.

Scalability and Flexibility: NXP's comprehensive portfolio allows OEMs to choose the specific components that best suit their vehicle designs and functionalities.

Future-Proof Designs: By adopting open standards and collaborating with industry leaders, NXP ensures that their solutions remain relevant as SDV technology evolves.

Looking Forward: NXP Paves the Way

NXP's commitment to innovation, robust hardware solutions, and fostering a collaborative ecosystem positions them as a key player in the SDV revolution. Their three pillars of support: silicon, analog components, and ecosystem collaboration, provide the essential foundation for building the next generation of software-defined vehicles. As the automotive industry continues to embrace software, NXP is well-positioned to play a leading role in shaping the future of mobility.

TEXAS INSTRUMENT

Texas Instruments: Gearing Up for the Software-Defined Vehicle Revolution

The automotive industry is undergoing a seismic shift with the rise of software-defined vehicles (SDVs). These vehicles prioritize software as the core, enabling features like over-the-air (OTA) updates, continuous improvement, and customization. As this revolution unfolds, Texas Instruments (TI) stands out as a key player, offering a comprehensive portfolio that empowers the development of next-generation SDVs.

A Legacy of Innovation: From Hardware to Software Enablement

TI has long been a dominant force in automotive electronics, supplying essential hardware components like microcontrollers and processors.

This established foundation provides them with a deep understanding of the unique needs and challenges of the car manufacturing process. However, recognizing the software-centric future of vehicles, TI is actively expanding its role in the SDV ecosystem.

TI's Portfolio: Building Blocks for the Software-Defined Future

TI's SDV-enabling portfolio encompasses a diverse range of products and solutions:

High-Performance Processors: TI's Sitara family of processors offers the necessary processing power to handle demanding software workloads in SDVs, including real-time data analysis for advanced driver-assistance systems (ADAS) and in-vehicle artificial intelligence (AI) applications.

Microcontrollers (MCUs): TI's broad portfolio of MCUs, including the Hercules and TMS570LC4x series, provides a range of options for various functionalities within an SDV. These MCUs are designed for automotive applications and prioritize safety and reliability.

Analog Hardware: Beyond processors and MCUs, TI offers a wide range of analog components that are crucial for efficient and reliable SDV operation. This includes power management solutions to optimize battery life in electric vehicles (EVs) and high-speed data converters for seamless communication between various electronic control units (ECUs).

Software Development Tools: TI's software development tools, like Code Composer Studio and TI RTOS, provide a robust platform for developers to create and optimize software for automotive applications.

Zone Architecture Support: The industry is moving towards a "zone architecture" for SDVs, where functionalities are consolidated into fewer, more powerful ECUs. TI's portfolio aligns with this trend, offering solutions that cater to the specific needs of central compute platforms, zone controllers, and edge modules within the vehicle.

The Benefits of TI's Approach

TI's comprehensive portfolio offers several benefits for companies developing SDVs:

Scalability and Flexibility: TI's diverse product range allows automakers to choose the most suitable hardware for their specific needs, ensuring scalability and flexibility in vehicle design.

Integration and Expertise: TI's long-standing experience in automotive electronics allows them to provide integrated solutions and technical expertise, streamlining the development process.

Focus on Safety and Security: TI prioritizes safety and security in its products, offering automotive-grade components that comply with stringent industry standards.

Looking Forward: Gearing Up for the Road Ahead

As the SDV revolution continues to gain momentum, TI is well-positioned to play a leading role. Their commitment to innovation, a diverse portfolio of hardware and software solutions, and focus on safety and security make them a valuable partner for companies developing the next generation of vehicles. By continuously evolving and adapting their offerings, TI will remain at the forefront of this technological transformation, shaping the future of mobility with software-defined vehicles.

RENESAS

Renesas: Powering Innovation in Software-Defined Vehicles

The automotive industry is experiencing a revolution driven by software-defined vehicles (SDVs). These vehicles prioritize software as the core, enabling features like over-the-air updates, continuous improvement, and customization. In this dynamic landscape, Renesas Electronics Corporation, a leading global supplier of microcontrollers and other automotive semiconductor solutions, is playing a crucial role.

Renesas: A Legacy of Automotive Expertise

Renesas boasts a rich history in the automotive industry, supplying essential electronic components for decades. This deep understanding of automotive electronics positions them exceptionally well to contribute to the evolution of SDVs. Here's how Renesas is leveraging their expertise:

Renesas R-Car Series: This family of high-performance SoCs (System-on-Chips) provides the processing power and functionality necessary for SDVs. The R-Car series features integrated hardware accelerators for AI and graphics processing, enabling advanced features like driver-assistance systems and in-vehicle infotainment.

Functional Safety: Safety remains paramount in the automotive industry. Renesas prioritizes functional safety in their R-Car SoCs, adhering to rigorous automotive safety standards like ISO 26262. This ensures the reliable operation of critical vehicle functions.

Security by Design: Cybersecurity is another top concern in the software-defined world. Renesas incorporates security features throughout the design process of their SoCs, safeguarding vehicles from potential cyber threats.

Driving Innovation: The Renesas R-Car Software Development Kit (SKD)

Beyond hardware, Renesas recognizes the importance of a robust software development environment for SDVs. A key component in their strategy is the Renesas R-Car Software Development Kit (SKD):

Tightly Coupled with Upstream Linux Kernel: The R-Car SKD offers a significant advantage by being tightly coupled with the upstream Linux kernel. This means it leverages the vast functionality and ongoing development of the Linux ecosystem. Developers benefit from a familiar and well-established foundation for building automotive software.

Flexibility and Faster Deployment: This tight coupling with the Linux kernel translates to greater flexibility for developers. They can readily access and utilize the extensive open-source libraries and tools available within the Linux ecosystem. This facilitates faster integration of new features and functionalities into SDVs.

Reduced Development Time: The R-Car SKD includes pre-integrated components and drivers specifically designed for the R-Car SoCs. This reduces the development time for automotive software engineers, allowing them to focus on building innovative features and functionalities.

Renesas R-Car Consortium: Building a Collaborative Ecosystem

In addition to the R-Car SKD, Renesas fosters collaboration by establishing the R-Car Consortium, a global community of automotive technology companies. This consortium facilitates knowledge sharing, joint development efforts, and the creation of a comprehensive ecosystem for SDV development.

The Benefits of Renesas Solutions for SDVs

Renesas' offerings provide several advantages for companies developing SDVs:

Reduced Development Time: The R-Car series, SKD, and associated development tools can help streamline the development process, leading to faster time-to-market for new SDV features.

Scalability and Flexibility: The R-Car family offers a range of SoCs with varying processing power, allowing automakers to choose the best fit for their specific vehicle designs and functionalities. The R-Car SKD further enhances flexibility by leveraging the Linux ecosystem.

A Collaborative Development Environment: The R-Car Consortium and the R-Car SKD foster collaboration and simplify software development for SDVs.

Looking Forward: Renesas on the Road to the Future

Renesas' commitment to innovation, robust hardware solutions, a developer-friendly software development kit, and fostering a collaborative ecosystem positions them as a key player in the SDV revolution. Their R-Car SoCs, development tools, and focus on safety and security empower companies to develop reliable, feature-rich, and secure software-defined vehicles. As the automotive industry embraces software, Renesas is well-positioned to play a critical role in shaping the future of mobility.

ST MICROELECTRONICS

STMicroelectronics: Orchestrating the Software-Defined Vehicle Symphony with a Diverse Portfolio

The automotive industry is undergoing a metamorphosis, with software taking center stage. STMicroelectronics (ST), a global leader in semiconductor solutions, is playing a vital role in this transformation by offering a comprehensive portfolio that empowers the development of software-defined vehicles (SDVs). This essay explores ST's diverse offerings, including the powerful Stellar P6 chip, established families like STM32, and the recently announced ST10 telematics unit, all working in concert to drive the SDV revolution.

Stellar P6: The Powerhouse for Demanding Software Tasks

At the heart of ST's SDV portfolio lies the groundbreaking Stellar P6 microcontroller (MCU). This high-performance chip boasts several features that make it ideal for the complex software needs of SDVs:

Unparalleled Processing Power: The Stellar P6 delivers exceptional processing muscle, enabling it to handle demanding tasks like real-time data analysis for advanced driver-assistance systems (ADAS) and in-vehicle artificial intelligence (AI) applications.

Safety First: ST prioritizes functional safety in the Stellar P6. It adheres to rigorous automotive safety standards like ISO 26262, ensuring the reliable operation of critical vehicle systems in all conditions.

Security by Design: Cybersecurity is a top concern in the SDV world. ST integrates robust security features throughout the Stellar P6 design, safeguarding vehicles from potential cyber threats.

Beyond Stellar P6: A Richer Symphony with STM32 and ST10

While the Stellar P6 takes center stage, ST's SDV portfolio extends beyond this single offering:

STM32 Microcontroller Family: The well-established STM32 family of MCUs provides a diverse range of options for various functionalities within an SDV. These MCUs are known for their reliability, efficiency, and wide range of features, making them suitable for a variety of applications from body electronics to powertrain control.

ST10 Telematics Unit: The recently announced ST10 telematics unit is another key player in ST's SDV orchestra. This unit provides essential functionalities for connected car services, a critical aspect of SDVs. The ST10 facilitates features like over-the-air (OTA) updates, remote diagnostics, and real-time vehicle data collection, all crucial for maintaining and improving SDVs throughout their lifespan.

ST's Strategic Vision: Collaboration and Openness

ST understands that fostering a collaborative and open ecosystem is essential for the success of SDVs. Here's how their strategy reflects this vision:

Partnerships for Innovation: ST actively collaborates with other industry players, like their joint development project with CARIAD (Volkswagen Group's software unit) for a new automotive system-on-chip (SoC). This open approach fosters innovation and leverages the strengths of different companies.

Focus on Efficiency and Scalability: ST recognizes the need for efficient power management and scalable solutions in EVs and other SDVs. Their portfolio includes solutions for battery management, power conversion, and tiered options within the STM32 family to cater to varying vehicle needs.

Open Standards and Tools: ST prioritizes open standards and development tools, making it easier for companies of all sizes to integrate their solutions into SDVs. This fosters a more flexible development environment and accelerates innovation.

A Diverse Customer Profile: Benefiting from ST's SDV Symphony

ST's comprehensive portfolio caters to a wide range of players in the automotive industry:

Original Equipment Manufacturers (OEMs): Large automotive manufacturers can leverage the high-performance Stellar P6 SoCs and the rich functionalities of the STM32 family to develop feature-rich and secure SDVs.

Tier 1 Suppliers: Tier 1 suppliers who design and develop automotive components can integrate ST's solutions, like the ST10 telematics unit, to add connected car functionalities to their systems.

Emerging Players and Startups: Startups and innovative companies developing new SDV technologies can benefit from ST's scalable solutions (like various STM32 options) and open collaboration approach to integrate seamlessly into the SDV ecosystem.

Looking Forward: ST Leads the SDV Orchestra

STMicroelectronics' commitment to innovation, a diverse portfolio like the Stellar P6, STM32 family, and ST10 unit, a focus on collaboration, and an open ecosystem approach position them as a key conductor in the SDV revolution. By empowering companies of all sizes to develop and integrate essential components, ST is at the forefront of shaping the future of software-defined vehicles. As the industry embraces software, ST is well-positioned to orchestrate a harmonious and innovative SDV future.

DENSO

Denso: Powering the Software-Defined Vehicle Revolution

The automotive industry is undergoing a seismic shift. Cars are no longer just about horsepower and fuel efficiency; they're rapidly evolving into sophisticated computers on wheels. This transition towards Software-Defined Vehicles (SDVs) presents both challenges and opportunities, and Denso, a global automotive technology leader, is at the forefront of this transformation.

From Hardware to Hardware + Software:

Denso, traditionally known for its prowess in automotive components like engines and thermal systems, is strategically adapting to this software-centric future. They recognize that while strong hardware remains crucial, software is becoming the central nervous system of modern vehicles.

Here's how Denso is approaching the SDV landscape:

Electronic Platform Development: Denso is actively developing robust electronic platforms that serve as the foundation for software applications in vehicles. These platforms will integrate hardware components with software functionalities, allowing for better communication and data exchange.

Focus on Advanced Driver-Assistance Systems (ADAS): ADAS features like automatic emergency braking, lane departure warning, and adaptive cruise control are becoming increasingly important. Denso is continuously refining these systems, aiming to develop more advanced functionalities that pave the way for autonomous driving in the future.

Connectivity and In-Car Experience: Denso understands the growing demand for seamless connectivity and a personalized in-car experience. They are developing software solutions that offer features like voice control, smartphone integration, and over-the-air software updates, ensuring vehicles stay up-to-date and cater to evolving user needs.

Focus on Future Technologies: Denso is actively exploring cutting-edge technologies like artificial intelligence and machine learning to further enhance SDVs. These technologies can power features like self-parking, predictive maintenance, and even personalized driving experiences.

Challenges and Collaboration

The transition to SDVs presents several challenges. Security vulnerabilities in software can pose significant safety risks.

Additionally, ensuring a robust and adaptable software architecture that can accommodate future advancements is crucial. To address these challenges, Denso is fostering collaboration with other industry players, including:

Strategic Partnerships: Collaborating with leading chipmakers to develop custom-designed chips specifically optimized for SDVs.

Open Innovation: Partnering with tech startups and software companies to leverage their expertise and accelerate innovation.

Looking Ahead: A Future Powered by Software

Denso's commitment to software development positions them as a key player in the SDV revolution. By harnessing their expertise in hardware and fostering collaboration, Denso aims to create safer, smarter, and more personalized driving experiences for the future. As software continues to reshape the automotive landscape, Denso is well-positioned to drive innovation and shape the future of mobility.

QNX

Powering the Future of Mobility: How QNX and Neutrino Real-Time Operating Systems Shape Software-Defined Vehicles

The automotive industry is undergoing a digital revolution. Cars are morphing from mechanical marvels into sophisticated computers on wheels, with software playing a central role in defining their functionality and performance. This shift towards Software-Defined Vehicles (SDVs) presents exciting opportunities.

Rock-Solid Foundations: QNX and Neutrino

At the heart of this software-driven automotive landscape lie two powerful operating systems: QNX by BlackBerry and Neutrino Real-

Time Operating System (RTOS). Here's why these technologies stand out as essential tools for building the future of mobility:

QNX: Security and Scalability:
Neutrino RTOS: Real-Time Performance:

Unlocking the Benefits of SDVs: A Combined Approach

By leveraging both QNX and Neutrino RTOS, car manufacturers can unlock a multitude of benefits for themselves and their customers:

Faster Innovation: Both QNX and Neutrino facilitate rapid development cycles, allowing manufacturers to bring new features and functionalities to market quicker.

Enhanced User Experience: QNX helps create a more personalized and engaging in-car experience through features like voice control, connected services, and over-the-air software updates.

Improved Safety: The robust security features of QNX and the real-time performance of Neutrino contribute to safer vehicles, reducing the risk of software-related malfunctions.

Unique Features for a Competitive Edge:

Both QNX and Neutrino offer unique features that set them apart:

QNX:Safety Certification: QNX is certified to meet rigorous automotive safety standards like ISO 26262, a crucial aspect for safety-critical applications in ADAS and autonomous vehicles.Hypervisor Support: QNX integrates seamlessly with hypervisors, enabling secure virtualization of different operating systems on a single hardware platform, further enhancing functionality and flexibility.

Neutrino RTOS:Microkernel Architecture: Similar to QNX, Neutrino's microkernel architecture promotes stability and security by isolating applications from the core operating system.Rich Connectivity Options: Neutrino supports various communication protocols, enabling seamless integration with different automotive components and sensors.

The Road Ahead: A Future Defined by Software

The automotive industry is rapidly evolving, and QNX and Neutrino RTOS are well-positioned to be driving forces in this transformation.

As software continues to define the driving experience, the power of QNX and Neutrino RTOS, can empower car manufacturers to deliver safer, smarter, and more personalized mobility solutions for the future.

VECTOR

The automotive landscape is undergoing a dramatic transformation. Cars are evolving beyond horsepower and fuel efficiency, morphing into complex computer networks – Software-Defined Vehicles (SDVs). This shift presents exciting possibilities, and Vector, a global leader in automotive software solutions, is at the forefront, providing the tools and expertise to make SDVs a reality.

Vector's SDV Products: Building Blocks for the Future

Vector offers a comprehensive suite of products that empowers engineers throughout the SDV development lifecycle:

1. Vector Base Layer: The Foundation for Connected Applications

The Vector Base Layer serves as the bedrock for various automotive applications, both in-vehicle and cloud-based. It includes:

AUTOSAR Classic and Adaptive Support: Enables developers to create modular and reusable software components based on the industry-standard AUTOSAR architecture.

MICROSAR Connect for the Big Data Loop: Facilitates secure and reliable data exchange between the in-vehicle network and the cloud. This is crucial for functionalities like over-the-air software updates and real-time diagnostics.

2. Vector Development Tools: Architecting the Future

Building complex software for SDVs requires the right tools. Vector provides a comprehensive suite that streamlines the development process:

CANoe: This industry-leading virtual validation and testing platform allows safe and efficient simulation of complex vehicle scenarios. Engineers can test software functionality in a virtual environment before deployment in real vehicles.

AUTOSAR Development Tools: Vector provides a complete AUTOSAR development environment, streamlining the creation and integration of AUTOSAR-compliant software components.

Security Solutions: With the growing reliance on software, security becomes paramount. Vector offers solutions for secure boot, intrusion detection, and secure communication, protecting vehicles from cyberattacks.

3. Vector XL 4 Platform: Powering High-Performance Computing in Vehicles

The Vector XL 4 platform is a powerful hardware platform specifically designed for high-performance computing (HPC) applications in vehicles. This platform is ideal for running demanding software applications like advanced driver-assistance systems (ADAS) and autonomous driving functions.

4. Vector Informatik GmbH Software Products:

In addition to the above, Vector Informatik GmbH offers a range of software products that cater to specific needs within the SDV development process. These include:

Safety-critical software solutions for applications requiring the highest levels of reliability and safety.

Diagnostic tools for troubleshooting and maintaining complex SDV systems.

Connectivity solutions that enable seamless communication between vehicles, infrastructure, and the cloud.

A Roadmap to a Software-Defined Future

The automotive industry is on the cusp of a software revolution. Vector, with its robust product portfolio, development tools, valuable expertise, and comprehensive services, is well-positioned to be a key partner in enabling car manufacturers to build the next generation of intelligent and connected vehicles. As software continues to define the driving experience, Vector is paving the way for a safer, smarter, and more personalized future of mobility.

QNX

Powering the Future of Mobility: How QNX and Neutrino Real-Time Operating Systems Shape Software-Defined Vehicles

The automotive industry is undergoing a digital revolution. Cars are morphing from mechanical marvels into sophisticated computers on wheels, with software playing a central role in defining their functionality and performance. This shift towards Software-Defined Vehicles (SDVs) presents exciting opportunities.

Rock-Solid Foundations: QNX and Neutrino

At the heart of this software-driven automotive landscape lie two powerful operating systems: QNX by BlackBerry and Neutrino Real-Time Operating System (RTOS). Here's why these technologies stand out as essential tools for building the future of mobility:

QNX: Security and Scalability:
Neutrino RTOS: Real-Time Performance:

Unlocking the Benefits of SDVs: A Combined Approach

By leveraging both QNX and Neutrino RTOS, car manufacturers can unlock a multitude of benefits for themselves and their customers:

Faster Innovation: Both QNX and Neutrino facilitate rapid development cycles, allowing manufacturers to bring new features and functionalities to market quicker.

Enhanced User Experience: QNX helps create a more personalized and engaging in-car experience through features like voice control, connected services, and over-the-air software updates.

Improved Safety: The robust security features of QNX and the real-time performance of Neutrino contribute to safer vehicles, reducing the risk of software-related malfunctions.

Unique Features for a Competitive Edge:
Both QNX and Neutrino offer unique features that set them apart:

QNX:Safety Certification: QNX is certified to meet rigorous automotive safety standards like ISO 26262, a crucial aspect for safety-critical applications in ADAS and autonomous vehicles. Hypervisor Support: QNX integrates seamlessly with hypervisors, enabling secure virtualization of different operating systems on a single hardware platform, further enhancing functionality and flexibility.

Neutrino RTOS:Microkernel Architecture: Similar to QNX, Neutrino's microkernel architecture promotes stability and security by isolating applications from the core operating system.Rich Connectivity Options: Neutrino supports various communication protocols, enabling seamless integration with different automotive components and sensors.

The Road Ahead: A Future Defined by Software

The automotive industry is rapidly evolving, and QNX and Neutrino RTOS are well-positioned to be driving forces in this transformation.

As software continues to define the driving experience, the power of QNX and Neutrino RTOS, can empower car manufacturers to deliver safer, smarter, and more personalized mobility solutions for the future.

TWENTY-ONE

INTERVIEW WITH DR EDWARD TSE

My journey through China's evolving automotive and technology landscape has been shaped not only by industry experience but also by the wisdom and insights of remarkable thought leaders. Among them, Dr. Edward Tse has had a profound impact on the way I understand China's socio-economic dynamics, innovation culture, and strategic direction.

Since 2013, I have had the privilege of engaging in deep, thought-provoking discussions with Dr. Tse—conversations that have spanned several hours over the years and significantly influenced my thinking process. His clarity of thought, global perspective, and nuanced understanding of China have been both intellectually enriching and personally inspiring.

As a gesture of gratitude and deep respect, I chose to include this interview with Dr. Tse in my book. I am truly thankful that he graciously agreed to be a part of it.

Interview with Dr. Edward Tse

Founder & CEO, Gao Feng Advisory | Author of "China's Disruptors"

1. "China is moving at lightning speed in automotive innovation, especially in SDVs. What systemic or cultural factors do you believe give China a unique edge over Western nations in this transformation?"

Answer: That's a very important question. In my work, I often refer to a concept I call the **"three-layered duality"** of China's innovation ecosystem. At the top, you have strong, strategic government direction. The Chinese government doesn't just regulate—it sets the vision. For instance, the NEV push or the focus on smart cities and intelligent mobility—these are not vague goals, they are national mandates.

Then you have the entrepreneurial dynamism at the grassroots level. Chinese entrepreneurs are not afraid to disrupt—they move quickly, iterate fast, and are often bold in their ambitions. Startups, tech firms, and even traditional OEMs are embracing SDVs because they see software as the new core of competitiveness.

The third layer is what I call the "middle"—local governments, state-owned enterprises, and research institutions. They act as accelerators. They build pilot zones, invest in infrastructure, and serve as early customers or partners. This is a powerful combination of **top-down policy and bottom-up entrepreneurship**, which is rare globally. Western countries often have one or the other, but not both working in sync.

China's openness to change, combined with its ability to scale solutions quickly, gives it a unique edge. In the case of SDVs, that means a rapid shift from concept to citywide deployment within months—not years.

2. "Chinese EV and SDV players like BYD, Xpeng, and NIO are not just competing—they are defining the future of mobility. How has China built such deep integration between tech, software, and automotive?"

Answer: You're absolutely right—companies like BYD, Xpeng, and NIO are no longer catching up with the West; they are defining what the future of mobility looks like. What makes China unique is that it never viewed the car as an isolated product. From the outset of the New Energy Vehicle (NEV) revolution, Chinese players saw vehicles as connected digital platforms—living ecosystems that evolve with the user.

Take Xpeng: they didn't just add screens and voice assistants; they engineered vehicles to be intelligent, self-learning, and fully integrated with the cloud. NIO took it a step further by creating an entire lifestyle ecosystem around the car—battery swap stations, user clubs, AI copilots—all designed to create an ongoing relationship between the user and the brand.

This kind of deep tech-automotive integration is rooted in China's **"internet mindset."** Unlike in many Western countries, where tech, auto, and AI industries often operate in silos, China encourages **fusion across domains.** Internet giants, AI labs, chipmakers, and cloud platforms actively collaborate with automotive companies. There's a cultural openness here—not being picky about the paths to innovation, but rather being **driven to learn fast, experiment, and evolve.**

Historically, this didn't happen overnight. Since the reform era beginning in 1978, and especially after 2012 under Xi Jinping's leadership, China has operated on a **dual-track economic model**: state-owned enterprises build infrastructure, while private companies innovate aggressively. This hybrid approach created fertile ground for disruptive ideas—like Software-Defined Vehicles (SDVs)—to scale rapidly.

Also, China's **governance model is not entirely centralized** in innovation. While policies provide direction and incentives, there is ample room for local and private experimentation. The outcome is an entire **internet-born ecosystem** being married with automotive DNA—leading to a new species of vehicles that are digital-first, AI-native, and constantly evolving.

Other countries may have legacy advantages, but China's **lack of legacy** in traditional car-making turned out to be its greatest strength—it skipped steps and reimagined mobility from the ground up.

3. "In your view, how critical is software—cloud, AI, chips, OTA—in China's vision to leapfrog global OEMs and become the epicenter of next-gen vehicle development?"

Answer: It's absolutely critical—software is no longer just a feature; it's the entire **foundation** of the SDV future. In traditional auto manufacturing, the emphasis was on horsepower, torque, and engineering excellence. Those still matter, of course—but now, the ability to deliver **digital experiences** and evolve them post-sale is just as important.

China understands this well. Whether it's Baidu's autonomous driving platforms, Huawei's vehicle-cloud ecosystem, or Horizon Robotics' AI chips, you're seeing **layered investments in the software stack**. These aren't side projects. They are strategic bets aimed at long-term differentiation.

And let's not forget Over-the-Air (OTA) updates. OTA allows you to treat the car like a smartphone—pushing features, security patches, or even revenue-generating upgrades after the car is sold. This changes the economics completely. It moves the business from one-time product sales to **lifetime value models**.

In this context, China's capability to scale fast, its control over supply chains, and its rapidly maturing software ecosystem give it a major advantage over traditional OEMs who are still learning to become software companies.

4. "The Chinese government has supported SDV and EV industries with policy, infrastructure, and capital. How would you compare China's approach with that of the US or EU?"

Answer: The contrast is quite stark. When China identifies an industry as strategic—like New Energy Vehicles (NEVs) or Software-Defined Vehicles (SDVs)—the support isn't symbolic. It's systemic. You see a full-stack approach: **consumer subsidies, R&D funding, factory land grants, pilot city programs, public-private testbeds**, and even dedicated industrial policies. It's not just about encouraging innovation—it's about enabling it at scale.

What's particularly effective is the **synchronized execution across government layers.** Ministries, provincial bodies, and city governments often act in alignment, pushing unified agendas at astonishing speed. A great example is battery swapping—once the government prioritized it, infrastructure appeared in dozens of cities within months. That level of coordinated rollout is almost impossible in more fragmented regulatory environments.

In contrast, the **U.S. has incredible tech innovation capacity**, but coordination across federal and state levels can be inconsistent, and often policy lags behind innovation. The **EU has strong regulatory vision**, especially around climate and sustainability, but its progress is often slowed by consensus-driven processes and stringent **data protection and privacy norms**, which, while ethically commendable, can hinder rapid progress in areas like connected mobility.

Another important point: **Europe missed the internet wave**. Unlike China, which embraced digital platforms early on, Europe didn't develop native internet giants. That absence now echoes in the mobility space, where software-defined capabilities are increasingly critical.

China's approach—**a hybrid model that is state-enabled yet market-driven**—lets it operate with **startup agility and national scale.** That's a potent formula. It's not that other regions lack talent or ambition, but China's ability to blend policy, capital, and execution gives it a first-mover advantage in the SDV revolution.

5. "Can Chinese SDV startups succeed globally, or will geopolitical tensions limit their reach? What's your outlook on China's potential to export not just vehicles, but mobility software and platforms?"

Answer: Geopolitical dynamics will absolutely shape the trajectory of Chinese SDV players globally—there's no denying that. However, **commercial value is a powerful force**, and Chinese companies are becoming increasingly strategic in navigating the global landscape. Rather than a one-size-fits-all expansion, they're **picking markets carefully, forming local partnerships**, and focusing on **value-led innovation** rather than simply competing on cost.

We've already seen BYD emerge as a dominant player in **Southeast Asia, Latin America, and now parts of Europe.** But beyond vehicle exports, the real long-term opportunity lies in **digital exports**—the **mobility software, platforms, and AI stacks** that power the next generation of vehicles.

Navigation systems tailored to local geographies, AI-driven voice cockpits, battery management platforms, and autonomous driving algorithms—**these are China's next big exports.** They don't need shipping lanes; they need APIs and cloud access. This makes Chinese firms far more scalable and adaptable in global markets.

Much like Huawei did with telecom infrastructure, I see Chinese SDV players evolving into **white-label mobility solution providers**—supplying the digital "brains" for vehicles across the world, even under other brand names. Already, we're seeing **technology transfer from China to the world** in areas like EV platforms and intelligent driving—this trend will only accelerate, especially in **automotive software and AI.**

In the bigger picture, China may not only export products—but **help shape the standards, protocols, and platforms** that define global mobility in the software-defined era.

6. "Many global OEMs are still stuck in legacy architecture and internal battles between software and hardware teams. Do you think China's 'start-from-zero' advantage has helped leap past these internal conflicts?"

Answer: Yes, and I would even say that this is **one of China's biggest structural advantages**. Legacy OEMs—especially in Europe and North America—are often dealing with deeply entrenched processes, old architectures, and rigid organizational structures. Software and hardware teams often operate separately, and integration is slow.

In contrast, many of China's top SDV players—especially startups—have been built from the ground up with software-first thinking. They adopted **centralized computing, zonal architecture, and modular software** models from day one. That means they can add features faster, fix bugs quicker, and experiment with new services without overhauling the entire system.

This isn't just about code—it's about culture. These new companies have **cross-functional teams**, faster decision cycles, and a mindset focused on digital UX rather than just mechanical performance. That's a huge differentiator.

7. "If you had to advise global automotive leaders on what to learn from China's SDV playbook, what would be your top three takeaways?"

Answer: Sure. I would say:

1. **Think Ecosystem, Not Product.** Don't just build a car—build an ecosystem. Bring in cloud partners, AI labs, payment systems, content providers. The SDV is not a standalone object anymore—it's part of a user's digital lifestyle.
2. **Speed is Strategy.** Launch early, iterate fast, improve constantly. Perfection is the enemy of progress. The Chinese approach is about momentum—make progress visible, build feedback loops, and keep evolving.
3. **Control the Software Stack.** Whoever controls the in-car OS, the cloud integration, and the AI layer controls the customer relationship—and the monetization potential. Don't outsource your brain. Build or co-create it.

Global OEMs still have many strengths—manufacturing quality, brand trust, and distribution networks—but if they don't adapt to the software-first era, they risk becoming hardware vendors in someone else's digital empire.

8. "As you have seen my book and you know the intent I have behind the book, what is your advise to me for my next edition"

Answer: First of all, I want to sincerely commend you on the exceptional work you've done with this book. The way you've unpacked complex concepts like Software-Defined Vehicles (SDVs) and Generative AI—making them both accessible and thought-provoking—is truly impressive. You've struck a rare balance between technical insight and engaging narrative, which is not easy to achieve.

For your next edition, I have a few suggestions that could deepen its impact even further.

First, consider adding a **dedicated technical section**—a "deep dive" for engineers, developers, and domain experts. While your current narrative is excellent for decision-makers and general readers, a more technical lens would add value for those looking to understand the architecture, algorithms, and real-world engineering behind SDVs and AI.

Second, I'd strongly recommend tracking and featuring **evolving case studies**—highlighting how key OEMs, Tier-1s, and startups are progressing over the next 2–3 years. Real-world examples of product launches, software platform rollouts, or regulatory experiments (like autonomous driving zones) will give your readers grounded, time-stamped insights that keep the book alive and evolving.

Third, and perhaps most important—**begin to write from the lens of an "AI-first mindset."** The world is moving toward intelligent systems that not only automate but learn, adapt, and evolve. The advent of **AI agents, autonomous decision systems, and even humanoid robots** is reshaping every industry. Including how these AI-native constructs interact with SDVs and broader manufacturing systems will allow your next edition to **extrapolate the future.**

Your perspective is unique. I'd encourage you to boldly explore what the **intersection of mobility, software, and intelligence** could look like in 2030 and beyond. You're not just writing about a technology shift—you're helping people understand a paradigm shift. Keep building on that.

BIBLIOGRAPHY, REFERENCES & SUGGESTED READINGS

Software-Defined Vehicles (SDVs):

1. **"Software-Defined Vehicles (SDV) Dummies Guide"** by Arm
   - A comprehensive overview of SDV technologies and their impact on the automotive industry. Arm

2. **"Automotive Software Engineering: Principles, Processes, Methods, and Tools"** by Jörg Schäuffele and Thomas Zurawka
   - An in-depth exploration of software engineering practices tailored for automotive applications.Amazon

3. **"Automotive Embedded Systems Handbook"** edited by Nicolas Navet and Francoise Simonot-Lion
   - Discusses the design and implementation of embedded systems in vehicles.

4. **"Model-Based Development Applications"** in *Automotive Software Engineering* by Miroslaw Staron
   - Focuses on model-based approaches for developing automotive software.

5. **"Vehicle Software Development: Bridging the Gap between Software and Automotive Engineering"** by Mark A. Nichols
   - Explores methodologies to integrate software development within automotive engineering processes.

Artificial Intelligence in the Automotive Industry:

1. **"The Future of the Automotive Industry: The Disruptive Forces of AI, Data Analytics, and Digitization"** by Inma Martinez
   - Examines how AI and data analytics are revolutionizing car design and functionality. Amazon

2. **"Artificial Intelligence for Autonomous Networks"** by Mazin Gilbert
   - Discusses AI applications in creating self-managing automotive networks.

3. **"Deep Learning for Autonomous Vehicle Control: Algorithms, State-of-the-Art, and Future Prospects"** by Sampo Kuutti et al.
   - A scholarly article on the role of deep learning in vehicle autonomy.

4. **"AI in the Automotive Industry: Applications and Innovations"** by Bryan P. Reimer
   - Provides insights into various AI applications transforming the automotive sector.

5. **"Machine Learning and Data Science in the Automotive Industry: Transforming Data into Intelligent Solutions"** by Patrick Bangert
   - Explores the intersection of machine learning and automotive data analytics.

Generative AI:

1. **"The Books You Need to Read to Master Generative AI"** by Ivo Bernardo
   - A curated list of essential readings to understand generative AI. Medium

2. **"Generative Deep Learning: Teaching Machines to Paint, Write, Compose, and Play"** by David Foster
   - A practical guide to generative models and their creative applications.

3. **"Hands-On Generative Adversarial Networks with PyTorch 1.x"** by John Hany and Greg Walters
   - Focuses on building and training GANs using PyTorch.

4. **"GANs in Action: Deep Learning with Generative Adversarial Networks"** by Jakub Langr and Vladimir Bok
   - Provides a hands-on approach to understanding and implementing GANs.

5. **"Deep Generative Models"** by Jakub M. Tomczak and Max Welling
   - Discusses the theory and application of deep generative models.

Electric Vehicles (EVs):

1. **"Electric Vehicles: Theory and Design"** by Yiqing Yuan
   - Offers a comprehensive understanding of EV design principles. Amazon

2. **"Electric Vehicle Technology Explained"** by James Larminie and John Lowry
   - An accessible introduction to the technology underpinning electric vehicles.

3. **"The Electric Vehicle Conversion Handbook"** by Mark Warner
   - A guide to converting traditional vehicles to electric power.

4. **"Electric and Hybrid Vehicles: Design Fundamentals"** by Iqbal Husain
   - Covers the fundamental concepts of electric and hybrid vehicle design.

5. **"Build Your Own Electric Vehicle"** by Seth Leitman and Bob Brant
   - A step-by-step guide to constructing an electric car.

Chinese Automotive Industry:

1. **"Decoding China's Car Industry: 40 Years"** by Li Anding
   - An authentic multidimensional history of China's car industry over four decades. Amazon

2. **"The Rise of China's Auto Industry and U.S.-Chinese Motor Vehicle Trade"** by Samantha Hutchins
   - Examines the growth of China's automotive sector and its trade relations with the U.S. Barnes & Noble

3. **"China's Disruptors: How Alibaba, Xiaomi, Tencent, and Other Companies Are Changing the Rules of Business"** by Edward Tse
   - Discusses how Chinese companies are reshaping global industries, including automotive.

4. **"The Chinese Automotive Industry: A Study of the Emerging Market"** by Thilo Hanemann and Daniel H. Rosen
   - Analyzes the development and future prospects of China's automotive market.

5. **"China's Drive to Electric Vehicles with Autonomous Characteristics: A Strategy for Competitiveness and Security"** by Richard P. Suttmeier
   - Explores China's strategic approach to EVs and autonomous vehicles.

Future Mobility:

1. **"Faster, Smarter, Greener: The Future of the Car and Urban Mobility"** by Venkat Sumantran, Charles Fine, and David Gonsalvez
   - Envisions a new world of connected and intelligent mobility. MIT Press

2. **"The Future of Mobility: Scenarios for the United States in 2030"** by Johanna Zmud et al.
   - Presents potential scenarios for mobility evolution in the U.S.

3. **"Shared Mobility: Innovation for the Future"** by Adam Cohen and Susan Shaheen
   - Discusses the role of shared mobility in transforming transportation.

4. **"Smart Cities, Smart Mobility: Transforming the Way We Live and Work"** by Lukas Neckermann
   - Explores the future of transportation in the context of urban development, connectivity, and sustainability.

5. **"The Mobility Revolution in the Automotive Industry"** by Dr. Sebastian Wedeniwski
   - Covers how digitalization, electrification, and shared mobility are reshaping car manufacturing and ownership.

6. **"Reinventing the Automobile: Personal Urban Mobility for the 21st Century"** by William J. Mitchell, Christopher E. Borroni-Bird, and Lawrence D. Burns
   - A visionary perspective on how vehicles and cities need to evolve together.

7. **"The End of Driving: Transportation Systems and Public Policy Planning for Autonomous Vehicles"** by Bern Grush and John Niles
   - A strategic look at the public policy implications of autonomous and shared transportation.

8. **"Mobility-as-a-Service: The End of Car Ownership?"** by David A. Hensher
   - Discusses the shift from car ownership to access and service-based mobility models.

Automotive Software Architecture & Engineering:

1. **"Software Engineering for Automotive Systems"** by Jane Fedorowicz and Mark Shackel
   - A structured approach to software lifecycle in automotive product development.

2. **"Software Engineering for Embedded Systems: Methods, Practical Techniques, and Applications"** by Robert Oshana
   - A solid reference for automotive-grade embedded systems software development.

3. **"Designing Embedded Systems and the Internet of Things (IoT) with the ARM mbed"** by Perry Xiao
   - Offers IoT and embedded system design techniques applicable in modern SDVs.

4. **"Systems Engineering and Analysis of Electro-Mechanical Products"** by Cornelius T. Leondes
   - Discusses systems engineering principles for cyber-physical automotive platforms.

5. **"Architecture Patterns with Python: Enabling Test-Driven Development, Domain-Driven Design, and Event-Driven Microservices"** by Harry J.W. Percival and Bob Gregory
    - Introduces event-driven, modular architectures suitable for modern automotive platforms.

6. **"Clean Architecture: A Craftsman's Guide to Software Structure and Design"** by Robert C. Martin
    - Must-read for designing robust, maintainable automotive software platforms.

Autonomous Vehicles & Robotics:

1. **"Autonomous Driving: How the Driverless Revolution Will Change the World"** by Andreas Herrmann, Walter Brenner, and Rupert Stadler
    - Covers social, economic, and technical aspects of the driverless revolution.

2. **"Self-Driving Cars: The New Way Forward"** by Michael Fallon
    - Explains the real-world progress, ethics, and societal impact of AVs.

3. **"Robotics and AI: Understanding the Future"** by Nathan Clark
    - Good for readers wanting to understand the role of robotics and AI in automotive evolution.

4. **"Introduction to Autonomous Robots: Mechanisms, Sensors, Actuators, and Algorithms"** by Nikolaus Correll
    - A solid foundation for readers interested in vehicle autonomy and control.

5. **"Deep Learning for Self-Driving Cars: Develop Next-Generation Autonomous Vehicles Using Artificial Intelligence"** by Szymon Rusinkiewicz
   - Combines deep learning techniques with AV control systems.

Cyber-Physical Systems & Embedded AI:

1. **"Cyber-Physical Systems: Foundations, Principles and Applications"** edited by Houbing Song, Danda B. Rawat, Sabina Jeschke, and Christian Brecher
   - An essential textbook on the underlying tech powering SDVs and connected vehicles.

2. **"Real-Time Concepts for Embedded Systems"** by Qing Li and Caroline Yao
   - Important for understanding RTOS and real-time considerations in vehicles.

3. **"Internet of Vehicles: Technologies and Services Toward Smart Cities"** by Rani, Lakshmi, et al.
   - Explores how vehicles will be part of smart city ecosystems.

4. **"Embedded Artificial Intelligence and Internet of Things"** edited by Anand Nayyar and Akshi Kumar
   - Discusses edge AI and embedded intelligence in automotive systems.

5. **"From Machine-to-Machine to the Internet of Things: Introduction to a New Age of Intelligence"** by Jan Holler et al.
   - Early but insightful read on M2M technologies in connected cars.

6. **"The Art of Designing Embedded Systems"** by Jack Ganssle
   - A practical guide for engineers working with automotive-grade embedded devices.

AI Agents & Humanoids:

1. **"Mastering AI Humanoid Robotics: Driving Generative AI Applications"** by Nicholas A. Smith
   - Explores the integration of generative AI into humanoid robotics, offering insights into design and application. Amazon

2. **"Humanoid Robotics: A Reference"** edited by Antonio Bicchi, Henrich Christoph, and Yoshihiko Nakamura
   - Comprehensive reference on the design, control, and application of humanoid robots.

3. **"Cognitive Robotics"** by Angelo Cangelosi and Minoru Asada
   - Discusses the intersection of cognitive science and robotics, focusing on learning and adaptation in humanoid robots.

4. **"Robot Ethics 2.0: From Autonomous Cars to Artificial Intelligence"** edited by Patrick Lin, Ryan Jenkins, and Keith Abney
   - Examines ethical considerations in robotics and AI, including humanoid applications.

5. **"Artificial Intelligence for Robotics: Build Intelligent Robots that Perform Human Tasks Using AI Techniques"** by Francis X. Govers
   - Practical guide on implementing AI techniques in robotics to perform human-like tasks.

6. **"Developmental Robotics: From Babies to Robots"** by Angelo Cangelosi and Matthew Schlesinger
   - Explores how principles of human development can be applied to robotics.

7. **"Humanoid Robotics and Neuroscience: Science, Engineering and Society"** edited by Gordon Cheng
   - Bridges the gap between neuroscience and humanoid robotics, discussing mutual insights.

8. **"Introduction to Autonomous Robots: Mechanisms, Sensors, Actuators, and Algorithms"** by Nikolaus Correll, Bradley Hayes, and Bradley M. Knoll
   - Provides foundational knowledge for building autonomous humanoid robots.

9. **"Human-Robot Interaction: An Introduction"** by Christoph Bartneck, Tony Belpaeme, Friederike Eyssel, Takayuki Kanda, Merel Keijsers, and Selma Šabanović
   - Introduces concepts and methodologies for studying interactions between humans and humanoid robots.

10. **"Robotics, Vision and Control: Fundamental Algorithms in MATLAB"** by Peter Corke
    - Offers algorithms and practical examples for vision-based control in robotics.

Connected Car Ecosystems:

1. **"Connected Vehicle Systems: Communication, Data, and Control"** by Bechara D. Saab and A. Wayne Johnson
   - Discusses recent advances in theory and practice of connected vehicle systems. Amazon

2. **"Connected Vehicles: Intelligent Transportation Systems"** by Radovan Miucic
   - Explores the technologies and applications of connected vehicles in intelligent transportation systems. Amazon+1SpringerLink+1

3. **"Internet of Vehicles: From Intelligent Grid to Autonomous Cars"** by Lu Yan, Yan Zhang, Laurence T. Yang, and Huansheng Ning
   - Covers the evolution of IoV and its role in autonomous driving.

4. **"Vehicular Ad Hoc Networks: Standards, Solutions, and Research"** by Claudia Campolo, Antonella Molinaro, and Riccardo Scopigno
   - – Provides insights into the standards and research surrounding VANETs.SpringerLink

5. **"Connected Vehicles in the Internet of Things: Concepts, Technologies and Frameworks for the IoV"** edited by Zaigham Mahmood
   - Presents an overview of smart transportation systems and IoV connectivity frameworks. SpringerLink

6. **"Automotive Ethernet: The Definitive Guide"** by Kirsten Matheus and Thomas Königseder
   - Details the role of Ethernet in automotive networking and connected vehicles.

7. **"Vehicle-to-Vehicle and Vehicle-to-Infrastructure Communications: A Technical Approach"** by Fei Hu
   - Explores V2V and V2I communication technologies and their applications.

8. **"The Car Hacker's Handbook: A Guide for the Penetration Tester"** by Craig Smith
   - Provides insights into the security aspects of connected cars.

9. **"Connected Cars: Technologies, Issues, and Market Developments"** by Andreas Riener, Myounghoon Jeon, and Ignacio Alvarez
   - Discusses the technological and market developments in connected cars.

10. **"Automotive Cybersecurity: An Introduction"** by Shiho Kim
    - Introduces cybersecurity concepts relevant to connected vehicles.ResearchGate

Safety, Cybersecurity, and Functional Safety:

1. **"Functional Safety in Modern Mobility: ISO 26262 and Beyond"** by Sridhar Radhakrishnan
   - Delves into automotive functional safety implemented in advanced electronic systems. Notion Press+1Amazon+1

2. **"Automotive SPICE, Safety and Cybersecurity Integration"** by Andreas Riel and Wolfgang Damm
   - Discusses the integration of Automotive SPICE with functional safety and cybersecurity standards. ResearchGate

3. **"Cybersecurity for Commercial Vehicles"** by Gloria D'Anna
   - Addresses cybersecurity challenges specific to commercial vehicles.

4. **"Automotive Cybersecurity: Principles and Practice"** by Keegan McNamara
   - Provides a comprehensive overview of cybersecurity principles in the automotive context.

5. **"Safety Critical Systems Handbook: A Straightforward Guide to Functional Safety, IEC 61508 (2010 Edition) and Related Standards"** by David J. Smith and Kenneth G. L. Simpson
   - Offers guidance on functional safety standards applicable to automotive systems.Amazon

6. **"Automotive Functional Safety: A Practical Guide to ISO 26262"** by Peter Clarke

7. **"ISO 26262: Road Vehicles – Functional Safety"** (Official Standard, ISO)
   - The international standard for automotive functional safety; must-read for understanding development lifecycle safety.

8. **"ISO/SAE 21434: Road Vehicles – Cybersecurity Engineering"** (Official Standard)
   - Global standard focusing on cybersecurity risks within automotive development.

9. **"Introduction to ISO 26262: Functional Safety in Automotive Development"** by David Ward
   - Beginner-friendly explanation of ISO 26262 and how it applies across vehicle systems.

10. **"System Safety Engineering and Risk Assessment: A Practical Approach"** by Nicholas J. Bahr
    - Explores hazard analysis, safety design, and risk assessments applicable to automotive systems.

Generative AI, Ethics & Society

1. **"Rebooting AI: Building Artificial Intelligence We Can Trust"** by Gary Marcus and Ernest Davis
   - Explores the limitations and risks of AI, urging a more grounded and trustworthy approach.

2. **"Weapons of Math Destruction"** by Cathy O'Neil
   - A critical look at how algorithmic decision-making can go wrong in real-world systems.

3. **"The Alignment Problem: Machine Learning and Human Values"** by Brian Christian
   - Investigates how AI can be aligned with human values, crucial in automotive safety & autonomy.

4. **"Artificial Intelligence: A Guide for Thinking Humans"** by Melanie Mitchell
   - Broad, insightful read about AI's progress, limits, and societal implications.

5. **"The Coming Wave: Technology, Power, and the 21st Century's Greatest Dilemma"** by Mustafa Suleyman
   - One of the co-founders of DeepMind explores AI's control problem and its global consequences.

6. **"AI 2041: Ten Visions for Our Future"** by Kai-Fu Lee and Chen Qiufan
   - Fiction + analysis on how AI will shape every industry, including mobility, healthcare, and education.

7. **"You Look Like a Thing and I Love You"** by Janelle Shane
   - A fun and accessible look into how neural networks actually function and misfunction.

Emerging Tech & Future Mobility

1. **"Mobility After COVID-19: A New Normal"** by T. Campisi et al.
   - Captures how global disruptions reset expectations in urban and vehicular mobility.

2. **"Future Urban Mobility"** by Matthias Finger & Maxime Audouin
   - Discusses multi-modal transport, connected infrastructure, and autonomous vehicles in smart cities.

3. **"Designing Autonomous AI: Reimagining Humanity and Technology"** by Ayanna Howard
   - Human-centered design principles in robotics and AI applications, relevant for SDVs and beyond.

4. **"Intelligent Transport Systems: Technologies and Applications"** by Asier Perallos et al.
   - Explores ITS infrastructure, V2X communications, and real-time traffic management systems.

5. **"How to Build a Car"** by Adrian Newey
   - A fascinating memoir of Formula One engineering—blending hardware with software precision.

6. **"Digital Twin Driven Smart Design"** by Fei Tao & Qinglin Qi
   - Explores digital twin applications in smart product lifecycle and manufacturing systems.

OEM Case Studies & Thought Leadership

1. **"Toyota Way 2.0: 14 Management Principles from the World's Greatest Manufacturer"** by Jeffrey K. Liker
   - A classic study of Toyota's disciplined engineering & lean principles still shaping modern OEMs.

2. **"Elon Musk"** by Walter Isaacson
   - Captures Tesla's philosophy, software-centric development, and radical engineering decisions.

3. **"Driving the Future: Combating Climate Change with Cleaner, Smarter Cars"** by Margo T. Oge
   - Former EPA director discusses innovations from GM, Tesla, and the broader EV ecosystem.

4. **"Zero to One"** by Peter Thiel
   - Not automotive-specific, but highly influential in thinking about future-proof, first-principle innovation.

5. **"Autonomy: The Quest to Build the Driverless Car"** by Lawrence D. Burns & Christopher Shulgan
   - A detailed account of GM's and Google's journey toward self-driving cars.

6. **"Innovating with Impact: The Case for Software-Defined Everything in the Auto Industry"** – McKinsey Automotive Insights (Whitepaper)
   - Strategic view on SDVs, software modularity, and platform thinking.

7. **"Annual Reports, Tech Roadmaps & Vision Statements"** from major OEMs like **Tesla, BYD, Toyota, BMW, Mercedes-Benz, Hyundai, and VW**
   - These are goldmines for tracking trends in software-defined platforms, autonomy, and AI.